STUDY GUIDE
TEACHER EDITION

GLENCOE
Macmillan/McGraw-Hill

New York, New York   Columbus, Ohio   Mission Hills, California   Peoria, Illinois

MERRILL CHEMISTRY

A GLENCOE PROGRAM

Student Edition
Teacher Wraparound Edition
Laboratory Manual–Student Edition
Laboratory Manual–Teacher Edition
English/Spanish Glossary
Computer Test Bank
Videodisc Correlations
Transparency Package
Problems and Solutions Manual
Solving Problems in Chemistry
Study Guide–Student Edition
Mastering Concepts in Chemistry–Software

Teacher Resource Books:
   Enrichment
   Critical Thinking/Problem Solving
   Study Guide–Teacher Edition
   Lesson Plans
   Chemistry and Industry
   ChemActivities
   Vocabulary and Concept Review
   Transparency Masters
   Reteaching
   Applying Scientific Methods in Chemistry
   Evaluation

Copyright by the Glencoe Division of Macmillan/McGraw-Hill School Publishing Company, previously copyrighted by Merrill Publishing Company.

All rights reserved. Permission is granted to reproduce the material contained herein on the condition that such materials be reproduced only for classroom use, be provided to students without charge, and be used solely in conjunction with the *Merrill Chemistry* program. Any other reproduction, for sale or other use, is expressly prohibited.

Send all inquiries to:

GLENCOE DIVISION
Macmillan/McGraw-Hill
936 Eastwind Drive
Westerville, OH 43081

ISBN 0-02-827227-7

Printed in the United States of America.

3 4 5 6 7 8 9 0 — MAL — 02 01 00 99 98 97 96

Contents

Chapter/Section		Page
1	The Enterprise of Chemistry	1
2.1	Decision Making	3
2.2	Numerical Problem Solving	4
3.1	Classification of Matter	7
3.2	Changes in Properties	8
3.3	Energy	9
4.1	Early Atomic Models	11
4.2	Parts of the Atom	13
5.1	Modern Atomic Structure	15
5.2	Quantum Theory	17
5.3	Distributing Electrons	19
6.1	Developing the Periodic Table	21
6.2	Using the Periodic Table	23
7.1	Symbols and Formulas	25
7.2	Nomenclature	26
8.1	Factor-Label Method	27
8.2	Formula-Based Problems	29
9.1	Chemical Equations	33
9.2	Stoichiometry	34
10.1	Periodic Trends	35
10.2	Reaction Tendencies	36
11.1	Hydrogen and Main Group Metals	37
11.2	Nonmetals	38
11.3	Transition Metals	40
12.1	Bond Formation	41
12.2	Particle Sizes	43
13.1	Bonds in Space	45
13.2	Molecular Arrangements	47
14.1	Molecular Attraction	51
14.2	Coordination Chemistry	53
14.3	Chromatography	54
15.1	Pressure	55
15.2	Motion and Physical States	56
16.1	Crystal Structure	57
16.2	Special Structures	60
17.1	Changes of State	63
17.2	Special Properties	66
18.1	Variable Conditions	69
18.2	Additional Considerations of Gases	71
19.1	Avogadro's Principle	73
19.2	Gas Stoichiometry	75
20.1	Solutions	77
20.2	Colloids	79
21.1	Vapor Pressure Changes	81
21.2	Quantitative Changes	83
22.1	Reaction Rates	85
22.2	Chemical Equilibrium	88
23.1	Acids and Bases	91
23.2	Salts and Solutions	94
24.1	Water Equilibria	97
24.2	Titration	100
25.1	Oxidation and Reduction Processes	101
25.2	Balancing Redox Equations	103
26.1	Cells	105

26.2	Quantitative Electrochemistry	107
27.1	Introductory Thermodynamics	111
27.2	Driving Chemical Reactions	113
28.1	Nuclear Structure and Stability	115
28.2	Nuclear Applications	118
29.1	Hydrocarbons	119
29.2	Other Organic Compounds	122
30.1	Organic Reactions and Products	125
30.2	Biochemistry	127
	Reduced Answer Pages	129

TO THE TEACHER

Most chapters in *Merrill Chemistry* are organized into two or three numbered sections. For each of these sections, the **Study Guide** book contains Study Guide masters that review the major vocabulary, facts, and concepts of the section.

Every Study Guide worksheet uses a strategy appropriate to the material in the section to lead students through the text presentation. These worksheets are especially useful with students who have difficulty comprehending the text and those who have trouble identifying important information.

Name _____ Date _____ Class _____

Chapter 1
STUDY GUIDE

THE ENTERPRISE OF CHEMISTRY

Complete each sentence.

1. Matter is anything that has the property of _____.

2. The way that matter behaves is described by its _____.

3. The property possessed by all matter, which in the proper circumstances can be made to do work, is called _____.

4. A property of matter that shows itself as a resistance to any change in motion is called _____.

5. Potential energy depends upon the _____ of an object with respect to some reference point.

6. Kinetic energy is the energy possessed by an object because of its _____.

7. Energy that is transferred as electromagnetic waves is called _____ energy.

Answer each question.

8. What do chemists do?
   _____
   _____
   _____

9. What are the two ways energy can be transferred between objects?
   _____
   _____

10. What does the law of conservation of mass state?
    _____
    _____

11. What does the law of conservation of energy state?
    _____
    _____

12. What does the law of conservation of mass-energy state?
    _____
    _____

13. When are changes of energy to mass and mass to energy observable?

14. What is an intermediate?

15. Why would water used in the making of a consumer product be considered a raw material and not an intermediate?

16. What is a model?

17. What law does Einstein's equation $E = mc^2$ express?

18. Explain why pesticides did not succeed in permanently controlling insect populations.

19. State two harmful effects of pesticides.

Name _____ Date _____ Class _____

Chapter 2
STUDY GUIDE

2.1 DECISION MAKING

Answer each of the following.

1. What are the two requirements that must be met before the answer to a chemistry problem can be called correct?
   _____

2. In solving problems, what should you do if the problem includes irrelevant information?
   _____

3. Why should you carefully consider the units in which your answer will be expressed?
   _____

4. Why is it important to examine your calculated answer to a problem?
   _____

5. In the table below, list the seven problem-solving steps outlined in your text. Then, read the problem and describe how to apply each of the seven steps to solve the problem.

   Problem: Seven students are working to put together a student newsletter. There are six sheets of paper in each newsletter. Three students can each assemble two newsletters per minute. Three other students can each assemble three newsletters per minute. The remaining student can assemble four newsletters per minute. How many newsletters will the group assemble in one hour?

Steps	Application of steps
(1)	
(2)	
(3)	
(4)	
(5)	
(6)	
(7)	

Name _____ Date _____ Class _____

Chapter 2
STUDY GUIDE

2.2 NUMERICAL PROBLEM SOLVING

1. Distinguish between quantitative and qualitative descriptions.

   _____

2. Complete the table.

Quantity	Unit	Unit symbol
Electric current		
	meter	
		s
Mass		
	kelvin	
Amount of substance		
Luminous intensity		cd

3. List the following units from largest to smallest: meter, millimeter, kilometer, centimeter, picometer.

   _____

Match each quantity to be measured with the most appropriate SI unit.

a. centimeter
b. gram
c. kelvin
d. kilogram
e. meter
f. newton
g. second

____ 4. length of a swimming pool

____ 5. length of a pencil

____ 6. mass of a pencil

____ 7. mass of a bag of apples

____ 8. time between two heartbeats

____ 9. temperature of boiling water

____ 10. weight of a textbook

11. Each of five students used the same ruler to measure the length of the same pencil. These data resulted: 15.33 cm, 15.34 cm, 15.33 cm, 15.33 cm, 15.34 cm. The actual length of the pencil was 15.55 cm. Using these numbers as examples, distinguish between accuracy and precision.

   _____
   _____

12. For each group of digits indicated in the three numbers below, state whether the digits are significant. Then state the rule that applies.

    a b c   d e   f
    4500.60 0.000 799 220

    a. _____
    b. _____
    c. _____
    d. _____
    e. _____
    f. _____

13. A stack of books contains 10 books, each of which is determined, by a ruler graduated in centimeters, to be 25.0 cm long. How do these two quantities differ in terms of significant digits?

14. What is the percent error in a determination that yields a value of 8.38 g/cm³ as the density of copper? The literature value for this quantity is 8.92 g/cm³.

15. A student measures the melting point of ammonium acetate, $NH_4C_2H_3O_2$, as 117°C, but the literature value is 114°C. What is the percent error in the measurement?

Derive units that would be appropriate for each of the following situations.

16. A faucet trickles, even when it is shut off. How would you express the rate of water flow?

17. When you add ice cubes to a glass of water, the water's temperature decreases. How would you express the change in temperature caused by the addition of each ice cube?

18. Complete the following table of densities, using Table 2.3 in your text to determine the probable identity of X, Y, and Z.

Material	Mass (g)	Volume (cm³)	Density (g/cm³)	Identity
X	16.0	2.0		
Y	17.8	2.0		
Z	2.7	3.4		

Name _____ Date _____ Class _____

Chapter 3
STUDY GUIDE

3.1 CLASSIFICATION OF MATTER

Answer each of the following.

1. How does a material differ from a mixture?

2. Describe an iceberg afloat in an ocean, using the terms *phase*, *state*, and *system*.

3. Compare and contrast heterogeneous and homogeneous mixtures and give two examples of each.

4. Suppose a glass contains 2 cups of sugar dissolved in 4 cups of water. Classify this mixture. Which component is the solute? Which is the solvent?

5. Explain why sea water is classified as a solution.

6. What is a substance? How does it differ from a homogeneous mixture?

7. How are elements and compounds related?

8. Are compounds more or less ordered than heterogeneous mixtures?

9. Classify the following materials as heterogeneous mixtures, compounds, elements, or solutions: sugar, salt, water, gold, brass, wood, carbon, and air.

10. Classify the following substances as organic or inorganic: methane (CH_4), aluminum chloride ($AlCl_3$), ethanol (C_2H_5OH), benzene (C_6H_6), and copper (Cu).

Name _____ Date _____ Class _____

Chapter 3
STUDY GUIDE

3.2 Changes in Properties

Write T for true and F for false. If a statement is false, change the underlined word or phrase and write your correction on the blank.

_____ 1. A change in which the same substance remains after the change is called a <u>chemical</u> change.

_____ 2. Melting is a <u>physical</u> change.

_____ 3. Evaporating water from a saltwater solution is a <u>chemical</u> change.

_____ 4. A <u>gaseous</u> substance that forms from a solution is called a precipitate.

_____ 5. The reaction of iron with hydrochloric acid is a <u>chemical</u> change.

Use the figure to answer the following questions.

6. What is the solubility of sodium chloride at 30°C? _____

7. Which is more soluble at 90°C: KNO_3 or $NaClO_3$? _____

8. At what temperature are the solubilities of KNO_3 and KBr the same? _____

9. At what temperature will 200 g of $NaClO_3$ dissolve in 100 g of H_2O? _____

Identify each property below as either a chemical property or a physical property. For each physical property, state whether it is extensive or intensive.

10. color _____

11. width _____

12. melting point _____

13. density _____

14. resistance to an acid _____

15. conductivity _____

16. luster _____

17. flammability _____

18. weight _____

8 STUDY GUIDE Chapter 3

Chapter 3
STUDY GUIDE
3.3 ENERGY

Answer the following questions.

1. Recall that ordinary salt is a compound of sodium and chlorine. List several systems that can be defined in a shaker full of salt.

2. What is heat?

3. What is a joule?

4. How is the calorie related to the joule?

5. Contrast endothermic and exothermic reactions.

6. What is specific heat?

Solve the following problems.

7. How many joules are in 1.11 Calories?

8. An average baked potato contains 164 Calories. Calculate the energy value of the potato in joules.

9. How much heat is lost when a 4110-g metal bar whose specific heat is 0.2311 J/g·C° cools from 100.0°C to 20.0°C?

10. What is the specific heat of copper if a 105-g sample absorbs 15 200 J and the change in temperature is 377°C?

11. What is the final temperature of a system of iron and water if a piece of iron with a mass of 25.0 grams at a temperature of 75.0°C is dropped into an insulated container of water at 35.5°C? The mass of the water is 150.0 grams. The specific heat of iron is 0.449 J/g·C°, and the specific heat of water is 4.184 J/g·C°.

Chapter 4
STUDY GUIDE
4.1 EARLY ATOMIC MODELS

Complete the following sentences.

1. The _____ is an apparatus that helped scientists discover that atoms contain electrons.

2. Isotopes of an element have the same number of protons but different numbers of _____.

3. The number of protons in the nucleus of an atom is called the _____ of that element.

4. The total number of protons and neutrons in a nucleus is called the _____.

5. The spontaneous production of rays of particles and energy by unstable nuclei is called _____.

Match the names of the scientists to their discoveries.

a. Einstein
b. Proust
c. Becquerel
d. Marie and Pierre Curie
e. Geiger, Marsden, and Rutherford
f. Millikan
g. Thomson
h. Lavoisier
i. Dalton
j. Chadwick

___ 6. ability of radium to give off rays

___ 7. positively charged nucleus

___ 8. $E = mc^2$

___ 9. electron's charge-to-mass ratio

___ 10. electron's charge

___ 11. neutrons

___ 12. law of conservation of mass

___ 13. ability of uranium to expose photographic film

___ 14. law of multiple proportions

___ 15. law of definite proportions

16. Briefly state the following:

   a. law of definite proportions
   _____
   _____

   b. law of multiple proportions
   _____
   _____
   _____

17. Complete the following table of isotopes of oxygen.

Isotope	Protons	Neutrons	Mass number
oxygen-16	8		16
oxygen-17	8	9	
oxygen-18	8		18

18. What is wrong with the diagram of a cathode-ray tube? How should it look? Explain your answer.
_____
_____
_____

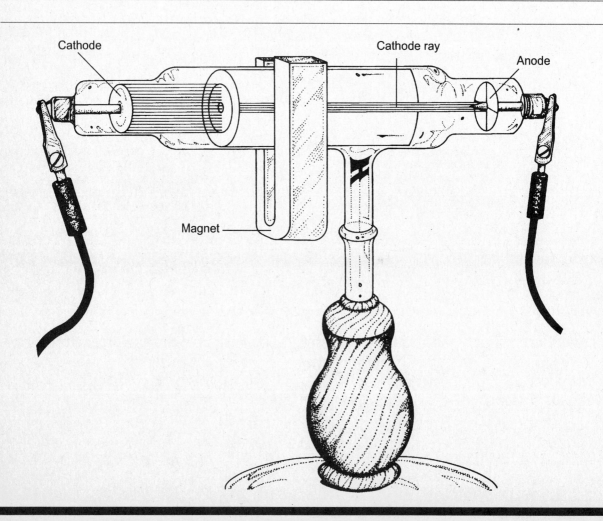

Name _____ Date _____ Class _____

Chapter 4
STUDY GUIDE

4.2 PARTS OF THE ATOM

Write T for true or F for false. If a statement is false, replace the underlined word or phrase with one that will make the sentence true, and write your correction on the blank provided.

_____ 1. Alpha radiation and beta radiation are made up of <u>particles</u>.

_____ 2. Gamma radiation is of very <u>low</u> energy.

_____ 3. Waves that are slightly higher in frequency than visible light are called <u>ultraviolet</u> rays.

_____ 4. Wavelength is represented by the Greek letter <u>nu (ν)</u>.

_____ 5. Electromagnetic energy travels at the speed of <u>sound</u>.

_____ 6. Electromagnetic waves with low frequencies have <u>long</u> wavelengths.

_____ 7. <u>Absorption</u> spectra are produced by energy given off by excited gaseous atoms.

_____ 8. The Rutherford-Bohr model of the atom is sometimes called the <u>planetary</u> model.

_____ 9. Planck proposed the equation <u>$E = mc^2$</u>.

_____ 10. A <u>photon</u> is a quantum of radiant energy.

Match the particles to the descriptions or examples.

a. beta particle
b. photon
c. alpha particle
d. baryon
e. meson
f. lepton
g. quarks
h. nucleon
i. antiparticle
j. gluon

____ 11. particle exchanged by quarks

____ 12. helium nucleus

____ 13. an electron, neutrino, muon, or pion

____ 14. always contains a quark and an antiquark

____ 15. a positron, for example

____ 16. an electron released by a nucleus

____ 17. always contains three quarks

____ 18. energy packet of light

____ 19. any component of a nucleus

____ 20. "up," "down," "charm," "strange," "top," or "bottom"

Name _____ Date _____ Class _____

Complete the following sentences.

21. _____ includes visible light, radio waves, and infrared, ultraviolet, and X rays.

22. Alpha, beta, and gamma rays given off by nuclei are all forms of _____.

23. Scientists use a nuclide of the element _____ as the standard for the atomic mass scale.

24. The symbol for an atomic mass unit is _____.

25. The _____ atomic mass takes into account the differing masses and percent occurrences of isotopes of an element.

26. Inside a mass spectrometer, the paths of heavy particles are bent _____ than are the paths of lighter particles.

27. Calculate the average atomic mass of lithium, which occurs as two isotopes that have the following atomic masses and abundances in nature: 6.017 u, 7.30%; and 7.018 u, 92.70%.

Name _____ Date _____ Class _____

Chapter 5
STUDY GUIDE

5.1 MODERN ATOMIC STRUCTURE

Complete the sentence or answer the question.

1. Define or explain the following.

   a. wave-particle duality of nature _____
   _____

   b. Newtonian mechanics _____
   _____

   c. quantum mechanics _____
   _____

   d. the Heisenberg uncertainty principle _____
   _____

   e. quantum numbers _____
   _____

2. The product of mass and velocity of an object is called the _____.

3. The French scientist whose hypothesis on the wave nature of particles helped lead to the present-day theory of atomic structure was _____.

4. The region in which an electron travels around the nucleus of an atom is called a(n) _____.

5. Complete the table, using de Broglie's equation, $\lambda = \dfrac{h}{mv}$, for the wavelength λ of a particle of mass m and velocity v. Note: Planck's constant $h = 6.63 \times 10^{-34}$ kg·m²/s.

m (kg)	v (m/s)	λ (m)
0.005 00	9.10×10^6	
0.0220		1.51×10^{-34}
	5.10	1.13×10^{-5}

6. Explain what is meant by the *probability* of finding an electron at a point in space.
_____
_____
_____
_____

Chapter 5   STUDY GUIDE   15

Write T for true or F for false. If a statement is false, replace the underlined word or phrase with one that will make the statement true, and write your correction on the blank provided.

_____ 7. De Broglie used Einstein's relationship between matter and energy and <u>Schrödinger's wave equation</u> to develop his equation for wavelength of a particle.

_____ 8. To be able to give a full description of an electron, you would need to know two things: its present position and its <u>radius</u>.

_____ 9. It is <u>impossible</u> to know both the exact position and the exact momentum of an object at the same time.

_____ 10. One of the most important developments of <u>Schrödinger's wave equation</u> was the idea of quantum numbers, which describe the behavior of electrons.

_____ 11. In Schrödinger's wave equation, n can have only <u>negative</u> whole number values.

Match each mathematical expression with the correct description.

a. de Broglie's equation for the wavelength of a particle
b. Einstein's equation relating matter and energy
c. Planck's quantum equation
d. term for total energy in Schrödinger's wave equation

____ 12. $\lambda = \dfrac{h}{mv}$

____ 13. $\dfrac{2\pi^2 me^4}{h^2 n^2}$

____ 14. $E = hv$

____ 15. $E = mc^2$

Name _____ Date _____ Class _____

Chapter 5
STUDY GUIDE

5.2 QUANTUM THEORY

Complete the sentence or answer the question.

1. When an electron in a hydrogen atom moves from a higher to a lower energy state, the energy difference is emitted as a quantum of _____.

2. Define the four quantum numbers n, l, m, and s, explain what information is given by each, and describe the range of values each may take.

   _____
   _____
   _____
   _____
   _____
   _____
   _____
   _____

3. Orbitals of the same energy are said to be _____.

4. The space occupied by one pair of electrons is called a(n) _____.

5. What is the formula for calculating the maximum number of electrons that can occupy any energy level in an atom? _____

6. Complete the following table.

Energy level	Number of sublevels	Number of orbitals	Maximum number of electrons
1			
2			
3			
4			

7. State the *Pauli exclusion principle*.

   _____
   _____
   _____

8. Complete the following table.

Sublevel	Number of orbitals	Maximum number of electrons
s		
p		
d		
f		

9. Complete the electron configurations for the following atoms by drawing in the arrows indicating the electrons with appropriate spins for each orbital.

                                    1s   2s   2p

a. beryllium (atomic number 4)

b. carbon (atomic number 6)

c. fluorine (atomic number 9)

Write T *for true or* F *for false. If a statement is false, replace the underlined word with one that will make the statement true, and write your correction on the blank provided.*

_____ 10. If two electrons occupy the same orbital, they must have <u>opposite</u> spins.

_____ 11. The principal quantum number describes the energy level of an <u>electron</u> in an atom.

_____ 12. The Schrödinger wave equation is solvable for any <u>multielectron</u> system.

_____ 13. <u>Four</u> quantum numbers are required to describe completely an electron in an atom.

_____ 14. The sum of all electron clouds in any sublevel (or energy level) is a <u>tetrahedral</u> cloud.

Name _____ Date _____ Class _____

Chapter 5
STUDY GUIDE

5.3 Distributing Electrons

Complete the sentence or answer the question.

1. In using the "rule of thumb" arrow diagram to predict electron arrangements for most atoms in the ground state, what assumption must be made?
   _____
   _____

2. It is important to know how to find the electron arrangement of an atom in order to predict its _____.

3. The number of electrons used in writing the electron configuration of an atom of an element is equal to the element's _____.

4. Write the electron configurations for the following elements:
   a. lithium _____
   b. boron _____
   c. sodium _____
   d. sulfur _____
   e. calcium _____

5. The notation in which outer-energy-level electrons are indicated around the symbol of an element is referred to as a(n) _____.

6. Write the three steps for drawing electron dot diagrams.
   _____
   _____
   _____
   _____

7. Write the electron dot diagrams for the following elements:
   a. lithium   d. sulfur

   b. boron   e. calcium

   c. sodium

Chapter 5   STUDY GUIDE 19

Chapter 6
STUDY GUIDE
6.1 DEVELOPING THE PERIODIC TABLE

Fill in the blanks with appropriate terms.

1. The table below shows the way a scientist named _____ classified the elements into groups he called _____.

Symbol	Atomic mass	Symbol	Atomic mass
Ca	40.1	Cl	35.5
Ba	137.3	I	126.9
Ca-Ba Average	88.7	Cl-I Average	81.2
Sr	87.6	Br	79.9

2. The table below shows the classification of elements according to a principle called the _____, which was developed by _____.

1	2	3	4	5	6	7
Li	Be	B	C	N	O	F
Na	Mg	Al	Si	P	S	Cl

3. The elements in Mendeleev's table were arranged in order of increasing _____.

4. As a result of Henry Moseley's work, the modern periodic table is arranged according to increasing _____.

5. The atomic number of an element indicates the number of _____ in the nucleus.

Write T for true and F for false. If a statement is false, replace the underlined word with one that makes the statement true.

_____ 6. The electron configurations of hydrogen and helium are <u>not similar</u>, so each element is in a separate column of the periodic table.

_____ 7. The elements in columns 3 through 12 (IIIB through IIB) are called the <u>noble gases</u>.

_____ 8. Each time a new principal energy level is started, a new <u>row</u> in the periodic table begins.

_____ 9. The lanthanoid series contains the elements lanthanum through <u>ytterbium</u>.

_____ 10. In the periodic table, a horizontal row of elements is called a <u>period</u>.

Answer the following questions.

11. What is the pattern of placing electrons in energy sublevels for elements in the actinoid series?
    _____
    _____

12. Why are neon and helium placed in the same column in the periodic table?
    _____
    _____

13. Compare the ways Dobereiner and Newlands classified the elements.
    _____
    _____
    _____

14. Fill in the following table.

Sublevel type	Electron capacity
	2
p	
	10
f	14

Chapter 6
STUDY GUIDE
6.2 USING THE PERIODIC TABLE

Match each of the following elements with the element in the list that has the most similar chemical properties. Use the periodic table as a guide.

a. zinc (Zn)
b. helium (He)
c. potassium (K)
d. rhodium (Rh)
e. terbium (Tb)
f. vanadium (V)
g. phosphorus (P)
h. gold (Au)
i. indium (In)
j. chlorine (Cl)
k. rhenium (Re)
l. oxygen (O)
m. yttrium (Y)
n. silicon (Si)
o. radium (Ra)

___ 1. cobalt (Co)
___ 2. carbon (C)
___ 3. argon (Ar)
___ 4. cerium (Ce)
___ 5. lithium (Li)
___ 6. aluminum (Al)
___ 7. iodine (I)
___ 8. copper (Cu)
___ 9. manganese (Mn)
___ 10. niobium (Nb)
___ 11. calcium (Ca)
___ 12. scandium (Sc)
___ 13. mercury (Hg)
___ 14. selenium (Se)
___ 15. arsenic (As)

1 1A																	18 VIIIA
1 H	2 IIA											13 IIIA	14 IVA	15 VA	16 VIA	17 VIIA	2 He
3 Li	4 Be					8	9 VIIIB	10				5 B	6 C	7 N	8 O	9 F	10 Ne
11 Na	12 Mg	3 IIIB	4 IVB	5 VB	6 VIB	7 VIIB				11 IB	12 IIB	13 Al	14 Si	15 P	16 S	17 Cl	18 Ar
19 K	20 Ca	21 Sc	22 Ti	23 V	24 Cr	25 Mn	26 Fe	27 Co	28 Ni	29 Cu	30 Zn	31 Ga	32 Ge	33 As	34 Se	35 Br	36 Kr
37 Rb	38 Sr	39 Y	40 Zr	41 Nb	42 Mo	43 Tc	44 Ru	45 Rh	46 Pd	47 Ag	48 Cd	49 In	50 Sn	51 Sb	52 Te	53 I	54 Xe
55 Cs	56 Ba	71 Lu	72 Hf	73 Ta	74 W	75 Re	76 Os	77 Ir	78 Pt	79 Au	80 Hg	81 Tl	82 Pb	83 Bi	84 Po	85 At	86 Rn
87 Fr	88 Ra	103 Lr	104 Unq	105 Unp	106 Unh	107 Uns	108 Uno	109 Une									

LANTHANOID SERIES	57 La	58 Ce	59 Pr	60 Nd	61 Pm	62 Sm	63 Eu	64 Gd	65 Tb	66 Dy	67 Ho	68 Er	69 Tm	70 Yb
ACTINOID SERIES	89 Ac	90 Th	91 Pa	92 U	93 Np	94 Pu	95 Am	96 Cm	97 Bk	98 Cf	99 Es	100 Fm	101 Md	102 No

Name _____ Date _____ Class _____

Write a T for true or F for false. If the statement is false, replace the underlined word with a word that makes the statement true.

_____ 16. The electron configurations of all elements in Group 1 (IA) end in s^1.

_____ 17. In the electron configuration for the outer energy level of potassium, $4s^1$, the coefficient 4 indicates the group number.

_____ 18. Atoms with full outer energy levels are very reactive.

_____ 19. Elements such as silicon are called metalloids because they have properties of both metals and nonmetals.

_____ 20. Elements with three or fewer electrons in the outer energy level are usually metals.

Answer each of the following.

21. Describe the properties of nonmetals, and give examples.
_____
_____
_____

22. What is the main reason that atoms react with each other?
_____
_____
_____

23. What determines an atom's chemical properties?
_____
_____

24. State the octet rule.
_____
_____

25. Describe the general positioning of metals and nonmetals in the periodic table.
_____
_____

Name _____ Date _____ Class _____

Chapter 7
STUDY GUIDE

7.1 SYMBOLS AND FORMULAS

Answer each of the following.

1. What is the most common source for an element's name? What are three other possible sources for the name of an element?
   _____
   _____

2. Complete the table by filling in the missing name or chemical symbol of each element shown.

Element	Symbol
	K
gold	
phosphorus	

3. a. What element does the symbol $^{40}_{20}Ca$ represent? _____
   b. What is this element's mass number? _____
   c. How many protons, electrons, and neutrons does the element contain? _____
   d. What would the charge be if an atom of this element lost two electrons? _____

4. Complete the table by filling in the missing information.

Chemical formula	Names of elements in compound	Relative number of each atom
NH_3		
$C_{12}H_{22}O_{11}$		
ZnF_2		
Fe_2O_3		

Write T for true or F for false. If a statement is false, change the underlined word(s) to make the statement correct by writing the correct word(s) on the blank.

_____ 5. <u>Atoms</u> are electrically charged particles.

_____ 6. In the formula of an ionic compound, the <u>negative</u> ion is written first.

_____ 7. The charge on a <u>diatomic molecule</u> is called its oxidation number.

8. Write the formulas of the compounds that will be formed when the positive ions listed in the table below combine with each negative ion listed.

	NO_3^-	S^{2-}	OH^-	Cl^-
Al^{3+}				
NH_4^+				
Pb^{4+}				

Chapter 7
STUDY GUIDE
7.2 NOMENCLATURE

Match the terms listed with their descriptions. Write the correct letters on the lines provided.

___ 1. binary compounds
___ 2. hydrocarbons
___ 3. sulfur hexafluoride
___ 4. aluminum arsenate
___ 5. baking soda
___ 6. *meth-*
___ 7. copper(II) oxide
___ 8. common acids
___ 9. cyclopentane
___ 10. nitrogen

a. example of a binary compound formed from a metallic element with variable oxidation states
b. indicates one carbon atom
c. organic compounds composed solely of H and C
d. forms five different binary compounds with oxygen
e. common name of sodium hydrogen carbonate
f. has five carbon atoms linked in a ring
g. compounds that contain only two elements
h. common name of a compound containing six fluorine atoms
i. have names that do not normally follow the rules for naming compounds
j. demonstrates the rule that the name of a polyatomic ion does not end in *-ide*

11. Complete the table by filling in the missing names.

Compound formula	Name
KCl	
$ZnCO_3$	
C_4H_{10}	
$FeCl_3$	
N_2O_3	

Complete the sentence or answer the question.

12. What is the difference between a molecular formula and an empirical formula?
_____
_____

13. The formulas for ionic compounds are almost all _____ formulas.

14. The molecular formula of a compound is always a whole-number multiple of the _____ formula.

15. What is the empirical formula of the compound N_2O_4?
_____

16. How would you use symbols to designate three formula units of zinc sulfide?
_____

Name _____ Date _____ Class _____

Chapter 8
STUDY GUIDE

8.1 FACTOR-LABEL METHOD

Complete the sentence or answer the question.

1. Define the following.
   a. factor-label method _____
   _____

   b. scientific notation _____
   _____

2. A conversion factor is a ratio equivalent to _____.

3. What advantage does scientific notation provide?
   _____
   _____

4. List the rules for handling decimal places or significant digits in the following kinds of calculations.
   a. addition and subtraction
   _____
   _____

   b. multiplication and division
   _____
   _____

5. Complete the following conversion-factor table.

Unit given	Unit desired	Conversion factor
m	cm	
kg	g	
nm	m	
h	min	
cm^3	dm^3	
mm	m	
m^3	cm^3	

6. Convert each of the following numbers to scientific notation.
   a. 124 _____
   b. 0.000 02 _____
   c. 564.45 _____
   d. 0.1001 _____
   e. 2.0 _____
   f. 301.03 _____

7. Enter the number of significant digits for each of the numbers below.
   a. 2.234 _____
   b. 124.3 _____
   c. 0.000 430 _____
   d. 50 000 _____
   e. 3.141 592 654 _____
   f. 78 456.1010 _____

8. Using the factor-label method, perform the following conversions. Express your answers in the correct number of significant digits.
   a. 30.0 min to h

   b. 125.1 km/s to m/min

   c. 0.003 m to cm

   d. 55.0 km/h to cm/s

9. Perform the following calculations. Express your answers to the correct number of significant digits and in scientific notation.
   a. add: $2.01 \times 10^3 + 4.23 \times 10^{-1}$

   b. subtract: $7.2 \times 10^2 - 7.1 \times 10^3$

   c. multiply: $10.1 \times 10^2 \times 6.23 \times 10^{23}$

   d. divide $\dfrac{5 \times 10^{-3}}{9.0 \times 10^6}$

Chapter 8
STUDY GUIDE

8.2 FORMULA-BASED PROBLEMS

Match the term with the correct description.

a. molecular mass
b. Avogadro constant
c. molarity
d. empirical formula
e. formula mass
f. molar mass
g. percentage composition
h. molecular formula

___ 1. the sum of the atomic masses of all atoms in the formula unit of an ionic compound

___ 2. the simplest ratio of the elements in a compound

___ 3. the sum of all the atomic masses in a molecule

___ 4. the mass of 6.02×10^{23} molecules, atoms, ions, or formula units of a species

___ 5. the ratio between the moles of dissolved substance and the volume of solution in cubic decimeters

___ 6. shows the actual number of atoms in a molecule

___ 7. 6.02×10^{23}

___ 8. a statement of the relative mass each element contributes to the mass of a compound as a whole

Find the formula mass or molecular mass of the following compounds.

9. ammonia, NH_3

10. methane, CH_4

11. sodium hydrogen carbonate (baking soda), $NaHCO_3$

Using the factor-label method, determine how many moles are represented by 25.0 g of each of the compounds listed in questions 9 through 11.

12. ammonia, NH_3

13. methane, CH$_4$

14. sodium hydrogen carbonate, NaHCO$_3$

Using the factor-label method, determine the mass, in grams, of 2.50×10^{24} molecules or formula units of each of the compounds in questions 9 through 11.

15. ammonia, NH$_3$

16. methane, CH$_4$

17. sodium hydrogen carbonate, NaHCO$_3$

18. Complete the following table by filling in the correct values.

Concentration	Moles of solute	Volume of solution
10.0M		3.00 dm^3
0.062M	0.651 mol	
	6.3 mol	5.9 dm^3
6.0M		1.0 dm^3
	0.0555 mol	0.500 dm^3

Describe the preparation of each of the following solutions.

19. 1.00 dm^3 of 6.00M NaCl

20. 2.50 dm^3 of 0.100M BaCl$_2$

Find the percentage composition of the following compounds.

21. carbon dioxide, CO_2

22. sulfuric acid, H_2SO_4

23. hydrogen peroxide, H_2O_2

Calculate the empirical formulas of the following.

24. a compound that is 45.9% K, 16.5% N, and 37.6% O

25. a compound, 5.00 g of which contains 4.28 g C and 0.720 g H

26. a compound, a 100.0-g sample of which contains 11.2 g H and 88.8 g O

Calculate the molecular formulas of the compounds listed below, given the empirical formula and formula mass of each.

27. empirical formula = CH, formula mass = 78.1 u

28. empirical formula = NH_2, formula mass = 32.1 u

Chapter 9
STUDY GUIDE
9.1 CHEMICAL EQUATIONS

Complete each sentence.

1. The process by which one or more substances are changed into one or more different substances is called a(n) _____.

2. The starting substances in a chemical reaction are called _____.

3. The symbol written after a formula to indicate a solid is _____.

4. In a balanced chemical equation, the number of _____ of any given kind on the left side is equal to the number on the right side.

Match each reaction type with the correct description.

a. combustion
b. double displacement
c. synthesis
d. single displacement
e. decomposition

___ 5. The positive and negative portions of two compounds are interchanged.

___ 6. A compound burns, reacting with oxygen.

___ 7. Two or more substances combine to form one new substance.

___ 8. A substance breaks down to form simpler substances when energy is supplied.

___ 9. One element replaces another in a compound.

Write T for true or F for false. If a statement is false, replace the underlined word or phrase with one that will make the statement true, and write your correction on the blank provided.

_____ 10. The symbol (aq) after a formula indicates that the substance is <u>dissolved in water</u>.

_____ 11. A number written to the left of a formula in an equation is called a <u>subscript</u>.

_____ 12. <u>Carbon</u> and water are typical products in a combustion reaction.

_____ 13. The substances generally written on the right in chemical equations are called <u>products</u>.

_____ 14. In a chemical equation, the symbol that stands for *yields* is <u>a plus sign</u>.

_____ 15. The general form of a <u>double displacement</u> reaction is: element + compound → element + compound.

_____ 16. In balancing a chemical equation, it <u>is</u> permissible to change subscripts.

Chapter 9
STUDY GUIDE
9.2 STOICHIOMETRY

Complete each sentence.

1. The branch of chemistry that deals with the amounts of substances involved in chemical reactions is called _____.

2. In solving a mass-mass problem, one must convert the number of grams of the given substance to _____.

3. The actual amount of product divided by the theoretical amount and multiplied by 100 is called the _____.

4. A(n) _____ problem is one in which the amount of heat absorbed or released during a reaction is calculated from information on the number of grams.

Write T for true or F for false. If a statement is false, replace the underlined word or phrase with one that will make the statement true, and write your correction on the blank provided.

_____ 5. The coefficients in a balanced chemical equation show the correct ratio of <u>masses</u>.

_____ 6. In solving a mass-mass problem, it is <u>not necessary</u> to work with a balanced chemical equation.

_____ 7. In a reaction with a low percentage yield, the amount of product produced, compared with the amount that theoretically could have been produced, is <u>small</u>.

_____ 8. The heat of reaction is represented by the letter <u>q</u>.

_____ 9. The heat of reaction for a system giving off energy has a <u>positive</u> value.

_____ 10. To convert grams to moles, the number of grams is <u>multiplied</u> by a factor that has the units mol/g.

_____ 11. A chemical equation that shows an energy term on the left represents a reaction that <u>absorbs</u> energy.

Name _____ Date _____ Class _____

Chapter 10
STUDY GUIDE

10.1 PERIODIC TRENDS

Write a definition for each of the following terms.

1. atomic radius _____
   _____

2. noble gas configuration _____
   _____

Answer each question in complete sentences.

3. Why does an atom of sodium have a larger atomic radius than an atom of chlorine has, even though sodium has fewer electrons?
   _____
   _____

4. When dissolved in water, NaCl will carry an electric current, but when the NaCl is in solid form, it will not do so. Explain this fact.
   _____
   _____

5. If iron has an electron configuration of $1s^2 2s^2 2p^6 3s^2 3p^6 4s^2 3d^6$, how does it form ions with oxidation numbers of 2+ and 3+?
   _____
   _____
   _____

6. Predict the most likely oxidation number for atoms that have the following electron configurations.
   a. $1s^2 2s^2 2p^6 3s^2 3p^6 4s^2 3d^{10}$ _____
   b. $1s^2 2s^2 2p^6 3s^2 3p^5$ _____
   c. $1s^2 2s^2 2p^6 3s^2 3p^1$ _____
   d. $1s^2 2s^2 2p^6 3s^2 3p^6 4s^2 3d^{10} 4p^6 5s^1 4d^{10}$ _____

Chapter 10
STUDY GUIDE
10.2 REACTION TENDENCIES

Write a definition for each of the following terms.

1. ionization energy _____
   _____

2. first ionization energy _____
   _____

3. shielding effect _____
   _____

4. electron affinity _____
   _____

Answer the following question in a complete sentence.

5. State the four factors that affect ionization energy, and briefly describe the effect of each.
   _____
   _____
   _____
   _____
   _____

Complete each of the following statements.

6. The first ionization energy tends to _____ as atomic number increases in any horizontal row, or period.

7. A column, or group, will show a decrease in first ionization energy as atomic number _____.

8. Ionization energy is typically measured in units called _____.

9. Metals have a _____ first ionization energy, and nonmetals have a _____ first ionization energy.

10. Nonmetals have _____ electron affinities. Metals have _____ electron affinities. Moving down a column, the values of the elements' electron affinities tend to _____.

Chapter 11
STUDY GUIDE

11.1 HYDROGEN AND MAIN GROUP METALS

Write T for true or F for false. If a statement is false, replace the underlined word or phrase with one that will make the statement true, and write your correction on the blank provided.

_____ 1. The shielding effect causes large atoms to lose their electrons more readily than they might otherwise.

_____ 2. A hydride ion is a bare proton.

_____ 3. A catalyst is a substance that speeds up a reaction.

_____ 4. By gaining an electron, hydrogen attains the stable electron configuration of helium.

_____ 5. As atoms of the alkali metals increase in size, they lose their outermost electron more easily.

_____ 6. Baking soda is another name for sodium carbonate.

_____ 7. The thallium ion and the ammonium ion each have a charge of 1+.

_____ 8. The alkali metal that is insoluble in most other alkali metals is cesium.

_____ 9. The alkali metals make up Group 2 of the periodic table.

_____ 10. Lime is a calcium compound.

_____ 11. Magnesium is the most plentiful metal in Earth's crust.

_____ 12. Alkaline earth metals tend to form 2+ ions.

Match each element with the correct description.

a. sodium
b. francium
c. hydrogen
d. potassium
e. magnesium
f. calcium
g. aluminum
h. beryllium
i. lithium
j. radium

___ 13. the most active metal
___ 14. is used in making nonsparking tools
___ 15. forms a silicate used as a catalyst and in soapmaking
___ 16. found in large amounts in muscle and nerve tissue
___ 17. can form both positive and negative ions
___ 18. its ions are a major constituent of bone and affect release and absorption of hormones
___ 19. the largest alkaline earth metal atom
___ 20. has three outermost electrons
___ 21. is part of a chlorophyll molecule
___ 22. alkali metal that reacts most vigorously with nitrogen and that burns in air to form an oxide, rather than a peroxide

Chapter 11
STUDY GUIDE
11.2 NONMETALS

Complete each sentence.

1. When most of the elements in Group 14 (IVA) react, they tend to _____ electrons.

2. Different forms of the same element are called _____.

3. Dry ice is the name for solid _____.

4. The Haber process produces _____.

5. The transfer of genetic information from generation to generation involves a nitrogen-phosphorus organic compound called _____.

6. The oxide ion has a charge of _____.

7. The form of oxygen that has the formula O_3 is called _____.

8. When SO_3 is dissolved in water, the compound called _____ is formed.

9. Ions made up of long chains of sulfur atoms attached to the S^{2-} ion are called _____ ions.

10. Chlorine and bromine are members of the _____ family.

11. The least reactive elements make up a family called the _____.

12. A mixture of metals is called a(n) _____.

13. A mixture of copper and tin is called _____.

14. When the members of Group 17 (VIIA) react, they usually _____ electrons.

15. Going down Group 17 (VIIA) of the periodic table, there is a(n) _____ in the activity of the elements.

Match each element with its description. Some choices may serve as answers more than once.

a. argon
b. oxygen
c. xenon
d. lead
e. fluorine
f. carbon
g. selenium
h. silicon
i. phosphorus
j. nitrogen
k. helium

___ 16. the most reactive nonmetallic element
___ 17. the most plentiful element in Earth's crust
___ 18. its allotropes are graphite and diamond
___ 19. the second most plentiful element in Earth's crust
___ 20. a distinctly metallic element in Group 14 (IVA)
___ 21. used in light bulbs to protect the filament
___ 22. a constituent of all organic compounds
___ 23. makes up most of Earth's atmosphere

___ 24. the first noble gas to produce a compound
___ 25. a semiconductor used in transistors and computer chips
___ 26. produces solder when mixed with tin
___ 27. in its white form, ignites spontaneously in air
___ 28. was first discovered somewhere other than on Earth
___ 29. a metalloid member of Group 16 (VIA)
___ 30. the only element to exhibit catenation to a great extent

Write T for true or F for false. If a statement is false, replace the underlined word or phrase with one that will make the statement true, and write your correction on the blank provided.

_____ 31. Carbonic acid and cyanides are examples of <u>organic</u> compounds.

_____ 32. <u>Tin</u> is an example of a metal that has been known since prehistoric times.

_____ 33. Some bacteria convert atmospheric <u>oxygen</u> to compounds that can be used readily by plants to make amino acids.

_____ 34. Elemental phosphorus occurs as <u>P_2</u> molecules.

_____ 35. A Group 15 (VA) element that occurs in all oxidation states from 3– to 5+ is <u>nitrogen</u>.

_____ 36. <u>Argon</u> is an example of an element that has never been made to form compounds.

_____ 37. Elements in the <u>carbon</u> family have complete outer energy levels.

Name _____ Date _____ Class _____

Chapter 11
STUDY GUIDE

11.3 TRANSITION METALS

Complete each sentence.

1. Steel is an alloy of _____.
2. The highest-energy electrons of transition metals are in the _____ sublevel.
3. There is a total of _____ columns, or groups, of transition metals in the periodic table.
4. The CrO_4^{2-} ion is called the _____ ion.
5. In most reactions, chromium loses _____ of its electrons.
6. Iron is galvanized by being dipped into molten _____.
7. Rubies and emeralds get their color from _____ impurities.
8. _____ is an alloy of copper and zinc.
9. The highest-energy electrons of the inner transition elements are in the _____ sublevel.
10. The inner transition elements of Period 6 are called _____.
11. The inner transition elements of Period 7 are called _____.
12. The most stable ion for lanthanoids has a charge of _____.
13. The lanthanoid element _____ forms alloys with unusual conductivity and magnetic properties.
14. _____ is a highly toxic actinoid.
15. A transition metal that is found in many proteins in biological systems is _____.

Match each element with its use described below.

a. silver
b. molybdenum
c. cobalt
d. zinc
e. manganese
f. neodymium
g. chromium
h. curium
i. titanium
j. osmium

___ 16. forms a dioxide used as a white paint pigment

___ 17. is used in spark plugs

___ 18. is needed in the diet for proper functioning of the pancreas

___ 19. is used to harden pen points

___ 20. forms a dioxide used in batteries

___ 21. is used in coinage

___ 22. is used to plate steel to protect it from corrosion

___ 23. may be used in the future as the energy source in satellite nuclear generators

___ 24. has a radioactive isotope used in cancer treatment

___ 25. forms an oxide used in glass filters and lasers

Chapter 12
STUDY GUIDE

12.1 BOND FORMATION

1. Define electronegativity.

2. Electron affinity and electronegativity are both measures of an atom's attraction for electrons. What is the difference between them?

3. How does electronegativity vary as the atomic number of an element increases within the same period of the periodic table?

4. How is the strength of a bond between two elements in a molecule related to their electronegativities?

5. What is the difference between an ionic and a covalent bond?

6. How is the character of a bond (ionic or covalent) between two elements related to their electronegativities?

7. Referring to Table 12.1, Electronegativities, in your text, arrange the following compounds in order of increasing ionic character of their bonds: LiF, LiBr, KCl, KI.

8. Referring to Tables 12.1 and 12.3 in your text, classify each of the following bonds as either ionic (I) or covalent (C):

   ___ a. Al–O   ___ f. N–O
   ___ b. Al–S   ___ g. Na–S
   ___ c. Bi–Cl   ___ h. P–O
   ___ d. Bi–O   ___ i. S–O
   ___ e. C–Cl   ___ j. Ti–Br

9. What force holds the two ions together in an ionic bond?

10. What is the meaning of the oxidation number of an element that forms an ionic bond?

11. What is a molecule?

12. What causes the bond lengths and bond angles of a molecule to vary?

13. a. Why does a molecular compound absorb specific frequencies of infrared radiation?

    b. How can the absorption of this radiation be used to identify a compound?

14. Why are the bonding electrons in metallic bonding said to be delocalized?

15. What is one factor that determines the hardness of a metallic element?

16. Indicate whether each property listed below is characteristic of ionic (I), covalent (C), or metallic (M) bonding. More than one letter may be used for each answer.

    ___ a. Shape of solid can be changed by pounding.

    ___ b. Not electrically conducting in solid phase

    ___ c. Electrically conducting in all phases

    ___ d. High melting points

17. a. How are a molecule and a polyatomic ion alike?

    b. How are they different?

Chapter 12
STUDY GUIDE
12.2 PARTICLE SIZES

1. a. Which is larger, a metallic atom or its positive ion?

   _____

   b. Which is larger, a nonmetallic atom or its negative ion?

   _____

2. In an ionic crystal, how can the internuclear distance between two ions be calculated?

   _____
   _____

3. In a molecule, how can the approximate bond length between two atoms be calculated?

   _____
   _____

4. Use Table 12.5 in your textbook to calculate the expected bond length (in picometers) of the following covalent bonds:

   _____ a. P–O
   _____ b. N–O
   _____ c. N–N

5. Arrange the atomic radius, covalent radius, and ionic radius of a nonmetallic atom in the expected order of increasing size.

   _____

6. Under what circumstances might the actual order of these sizes change?

   _____
   _____
   _____

Chapter 13
STUDY GUIDE

13.1 BONDS IN SPACE

Complete the sentence or answer the question.

1. Pairs of electrons that bond two atoms in a molecule are called _____.

2. Why do electron pairs in the outer levels of atoms in a molecule spread apart as far as possible?
   _____

3. Because each of the bond angles of methane equals 109.5°, its molecular shape is a perfect _____.

4. Complete the following diagrams to show the electron content of the outer orbitals of (a) an unbonded carbon atom, and (b) a hybridized carbon atom.

   Unbonded carbon atom   Hybridized carbon atom
   Outer orbitals   Outer orbitals

   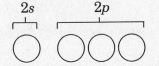 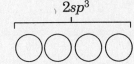

5. How do hybridized orbitals in an atom compare with one another?
   _____
   _____

6. When an sp^3 orbital of one carbon atom overlaps that of another carbon atom, how many electrons do the two atoms share?
   _____

7. How many hybrid orbitals of the following types can an atom have?
   a. sp^3 ___
   b. sp^2 ___
   c. sp ___

8. In a molecule of ethylene, C_2H_2, how many sigma bonds and how many pi bonds are present?
   _____

9. In a trigonal planar structure such as the CH_2O molecule, with one double bond and two single bonds, how does the double bond affect the angle between the two single bonds?
   _____
   _____

Write T *for true or* F *for false. If a statement is false, replace the underlined word or phrase with one that will make the statement true, and write your correction on the blank provided.*

_____ 10. A <u>shared pair</u> of electrons is formed when an orbital of one atom overlaps an orbital of another atom.

_____ 11. <u>Sigma</u> bonds are always formed by the sideways or parallel overlap of unhybridized *p* orbitals.

_____ 12. In single, double, and triple bonds between carbon atoms, one of the bonds is always a <u>pi</u> bond.

_____ 13. When two carbon atoms are joined by more than one bond, the additional bonds are always <u>pi</u> bonds.

_____ 14. Pi bonds break more easily than do sigma bonds, because the electrons forming pi bonds are <u>closer to</u> the nuclei.

Chapter 13
STUDY GUIDE
13.2 MOLECULAR ARRANGEMENTS

Complete the sentence or answer the question.

1. In saturated organic compounds, all the bonds between carbon atoms are _____.

2. The existence of two or more substances with the same molecular formula but different structural formulas is known as _____.

3. What is the difference between structural isomerism and positional isomerism?
   _____
   _____

4. Compounds with the same atoms bonded in the same order but with a different arrangement of atoms around a double bond are examples of _____.

5. a. What type of hybridization would be expected in an atom with three outer electrons?
   _____

   b. What would be the shape and bond angles of the molecule it forms?
   _____

Match each suffix with its correct description of the bonding of the carbon atoms in a hydrocarbon.

a. saturated compounds (single bonds only)
b. unsaturated compounds with double bonds
c. unsaturated compounds with triple bonds

___ 6. -yne

___ 7. -ane

___ 8. -ene

Match each of the following compounds with the correct description of its molecule.

a. boron trichloride (BCl_3)
b. acetylene (ethyne) (C_2H_2)
c. ozone (O_3)
d. oxygen difluoride (OF_2)
e. carbon tetrachloride (CCl_4)
f. trimethylarsine $(CH_3)_3As$

___ 9. linear, with 180° bond angles

___ 10. bent, with an exceptionally small bond angle resulting from two unshared pairs of electrons

___ 11. bent, with a small bond angle resulting from one unshared pair of electrons

___ 12. tetrahedral, with the expected bond angles of 109.5°

___ 13. trigonal pyramidal, with an unusually small bond angle (96°)

___ 14. trigonal planar, with the expected bond angles of 120°

For each of the following pairs of isomers, describe the difference in structure and identify the type of isomerism illustrated.

15.

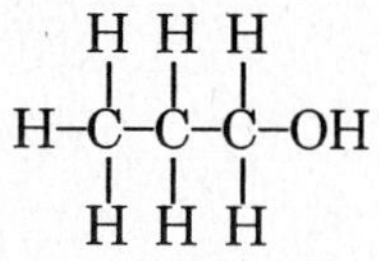

```
    H H H                          H OH H
    | | |                          |  |  |
  H-C-C-C-OH                     H-C - C - C-H
    | | |                          |  |  |
    H H H                          H  H  H

   1-propanol                     2-propanol
```

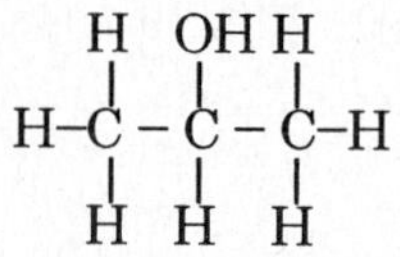

Structural difference: _____

Type of isomerism: _____

16.

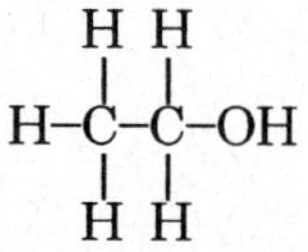

```
    H H                              H    H
    | |                              |    |
  H-C-C-OH                         H-C-O-C-H
    | |                              |    |
    H H                              H    H

   ethanol                        methoxymethane
```

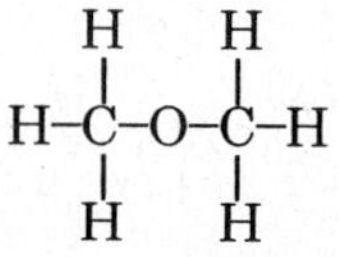

Structural difference: _____

Type of isomerism: _____

Complete the sentence or answer the question.

17. a. In the compound named ethene, what does the ending *-ene* indicate about its structure?

_____
_____

b. What does the stem *eth-* indicate?

_____
_____

18. a. Draw the structural formula for $CH \equiv C-CH_2-CH_3$.

b. Give the name of this compound, and explain the meaning of its stem and ending.

_____
_____
_____

48 STUDY GUIDE   Chapter 13

Name _____ Date _____ Class _____

19. Under each of the following simplified structural diagrams, write the name of the compound represented.

□   ⬠   ⬡(with double bond)   ⬢

_____   _____   _____   _____

20. Draw the structural formula for the positional isomer of the compound in question 18.

Chapter 13   STUDY GUIDE 49

Chapter 14
STUDY GUIDE

14.1 MOLECULAR ATTRACTION

Complete the sentence or answer the question.

1. Define the following terms.

   a. polar covalent bond _____

   b. dipole _____

   c. dipole moment _____

2. Are the polar bonds in a polar molecule arranged symmetrically or asymmetrically?
   _____

3. What does the dipole moment of a molecule tell us about its intermolecular forces?
   _____

4. Indicate whether the following molecules are polar or nonpolar.

   a. BeF_2 _____

   b. H_2O _____

   c. $CHCl_3$ _____

   d. CCl_4 _____

Write the letter of the term that matches the correct description.

a. van der Waals forces
b. intramolecular forces
c. intermolecular forces
d. dipole-dipole forces
e. dipole-induced dipole forces
f. temporary dipole
g. dispersion forces
h. induced dipole

___ 5. a nonpolar molecule in which the charge distribution is briefly asymmetrical

___ 6. a nonpolar molecule that has its electron cloud distorted by an approaching dipole and is thus transformed into a dipole

___ 7. weak forces involving the attraction of the electrons of one atom for the protons of another

___ 8. attractive forces between two molecules of the same or different substances that are both permanent dipoles

___ 9. forces generated by temporary dipoles

Name _____ Date _____ Class _____

___ 10. forces between molecules

___ 11. attractive forces between dipoles and nonpolar molecules

___ 12. forces within a molecule

Write T *for true and* F *for false. If a statement is false, replace the underlined word or phrase with one that will make the statement true, and write your correction on the blank provided.*

_____ 13. <u>Dispersion</u> forces are the only attractive forces that attract between nonpolar molecules.

_____ 14. Molecules <u>cannot</u> exhibit both dipole and dispersion interactions.

_____ 15. Substances composed of nonpolar molecules are generally gases at room temperature or <u>high-boiling liquids.</u>

_____ 16. Substances composed of polar molecules generally have <u>higher</u> boiling points than do nonpolar compounds.

_____ 17. A nonpolar molecule <u>can</u> have polar bonds.

_____ 18. Water is a <u>nonpolar</u> molecule that has polar bonds.

Name _____ Date _____ Class _____

Chapter 14
STUDY GUIDE

14.2 COORDINATION CHEMISTRY

Complete the sentence or answer the question.

1. What is a complex ion?
   _____

2. Name two uses for complex ions.
   _____

3. A ligand is either a _____ or a _____ that is attached to a central positive ion in a complex ion.

4. The coordination number of a complex ion is the number of _____ of ligands surrounding the central positive ion.

5. Why is the oxalate ion in a complex ion called a didentate ligand?
   _____

6. What type of complex is formed by three didentate ligands? _____

7. The molecules of a coordination compound are complexes with a net charge of _____.

Write T for true and F for false. If a statement is false, replace the underlined word or phrase with one that will make the statement true, and write your correction on the blank provided.

_____ 8. In a complex ion, <u>nonpolar</u> molecules or negative ions are attached to a central positive ion.

_____ 9. Ligands can be either molecules or <u>positive</u> ions.

_____ 10. The most common ligand is <u>water</u>.

_____ 11. Complex ions with coordination number <u>two</u> are always linear.

_____ 12. In a coordinate covalent bond the electrons in the shared pair come from <u>different atoms</u>.

_____ 13. Although the bonds of most complex ions have characteristics of both covalent and ionic bonding types, the <u>ionic</u> character dominates.

14. Complete the following table.

Coordination number	Shape of complex
2	
4	
6	

Name _____ Date _____ Class _____

Chapter 14
STUDY GUIDE

14.3 CHROMATOGRAPHY

Complete the sentence or answer the question.

1. Define the following terms.

    a. fractionation _____

    b. chromatography _____
    _____

2. Which phase in chromatography consists of the mixture to be separated being dissolved in a fluid (liquid or gas)? _____

3. Label the following parts of the gas chromatography system in the figure.

    a. inert carrier gas
    b. stationary phase
    c. mobile phase
    d. volatile mixture to be separated

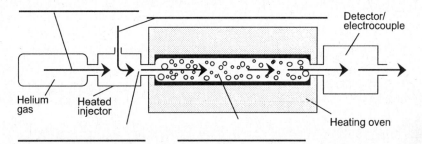

Write T for true or F for false. If a statement is false, replace the underlined word or phrase with one that will make the statement true, and write your correction on the blank provided.

_____ 4. Chromatography is a method of separating a mixture of substances into components based on differences in <u>polarity</u> of the components.

_____ 5. In chromatography, the fastest migrating substance will be the one with the <u>least</u> attraction for the stationary phase.

_____ 6. <u>Paper</u> chromatography is used when it is necessary to speed up the separation process.

_____ 7. An ion exchange resin is used as the stationary phase of <u>thin layer</u> chromatography.

Write the letter of the type of chromatographic technique that matches the description.

a. column chromatography
b. high performance liquid chromatography
c. ion chromatography
d. paper chromatography
e. thin layer chromatography
f. gas chromatography

___ 8. combines some of the techniques of both column and paper chromatography and is used frequently in separating biological materials

___ 9. is a chromatographic technique for the analysis of volatile liquids and mixtures of gases

___ 10. is designed to overcome the speed limitations of conventional column chromatography

___ 11. is a simple and fast chromatographic technique, but one in which quantitative determinations are difficult because of its extremely small scale

___ 12. is used for vitamins, proteins, and hormone separations not easily made by other methods

Name _____ Date _____ Class _____

Chapter 15
STUDY GUIDE
15.1 PRESSURE

Write T for true or F for false. Is a statement is false, replace the underlined word or phrase with one that will make the statement true, and write your correction in the blank.

_____ 1. All matter is composed of <u>small particles</u>.

_____ 2. Particles that make up matter are always <u>stationary</u>.

_____ 3. Collisions between particles of matter are perfectly elastic because there is <u>no change</u> in the total kinetic energy of the particles before and after the collision.

_____ 4. Under ordinary conditions, each molecule of a gas undergoes a few <u>million</u> collisions each second.

_____ 5. A barometer is <u>an open-arm</u> manometer used to measure atmospheric pressure.

_____ 6. Standard atmospheric pressure is <u>one pascal</u>.

Complete the statement or answer the question.

7. What is one pascal?
   _____

8. In using any manometer, what measurement must you make?
   _____

9. What physical property of the liquid in a manometer must you know?
   _____

10. Why is the measurement of gas pressure with a closed-arm manometer independent of the atmospheric pressure?
    _____

11. Standard atmospheric pressure will support a column of mercury _____ mm high.

12. One kilopascal equals _____ mm Hg.

13. A closed-arm mercury manometer is shown being used to measure the pressure of a gas in a closed container. What is the pressure of the gas in kilopascals?

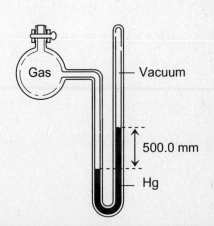

Chapter 15   STUDY GUIDE   55

Chapter 15
STUDY GUIDE
15.2 MOTION AND PHYSICAL STATES

Answer the following questions.

1. Complete the following table comparing the three common states of matter.

State of matter	Shape	Volume
solid		definite
	assumes the shape of its container	
gas		

2. What two factors determine the average speed of the particles in a gas?

3. Define *absolute zero*.

4. In which direction does energy flow between two objects at different temperatures?

5. What is the SI unit of temperature?

6. What is absolute zero on the Celsius temperature scale?

7. What equation can be used to convert Celsius temperatures (°C) to Kelvin temperatures (K)?

8. Convert the following temperatures to the Kelvin scale.
    a. 15°C _____
    b. −100°C _____
    c. 0°C _____

9. Convert the following temperatures to the Celsius scale.
    a. 0 K _____
    b. 100 K _____
    c. 722 K _____

10. What is the difference between heat and temperature?

Chapter 16
STUDY GUIDE
16.1 CRYSTAL STRUCTURE

1. Complete the table.

Crystal system	Lengths of unit cell axes	Angle between unit cell axes
cubic	equal	
tetragonal		all = 90°
	all unequal	all = 90°
	all unequal	2 = 90°, 1 ≠ 90°
triclinic		
rhombohedral		
hexagonal		1 = 90°, 3 = 60°

2. Identify each of the crystal shapes in Figure 1.

Figure 1

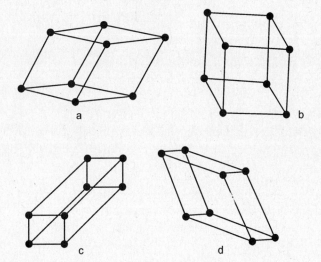

a. _____

b. _____

c. _____

d. _____

3. Identify each of the unit cells in Figure 2.

Figure 2

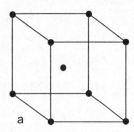 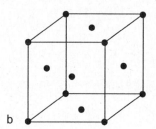

a.

b.

a. _____

b. _____

Figure 3

4. a. What is the general name for a structure like the one shown in Figure 3?

   _____

   b. If the unit cell is simple cubic, how many unit cells does the structure in Figure 3 contain?

   _____

Match each type of bonding with the correct type of crystal.
a. covalent
b. delocalized electrons
c. electrostatic forces
d. van der Waals forces

____ 5. ionic

____ 6. metallic

____ 7. macromolecular/network

____ 8. molecular

9. Label each range of melting points on the temperature scale in Figure 4 with the name of the appropriate crystal type.

Figure 4

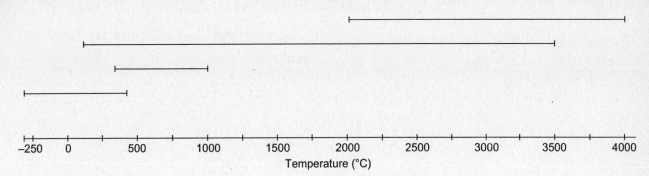

Write T *for true or* F *for false. If a statement is false, replace the underlined word or phrase with one that will make the statement true, and write your correction on the blank provided.*

_____ 10. In a crystal of sodium chloride, each Na⁺ ion is surrounded by <u>four</u> Cl⁻ ions.

_____ 11. The unit cell of sodium chloride is <u>body-centered</u> cubic.

_____ 12. Crystals of different solids with the same structure and shape are said to be <u>polymorphous</u>.

Chapter 16
STUDY GUIDE

16.2 SPECIAL STRUCTURES

1. Label the crystal defects shown in Figure 1. Explain how each defect is caused.

Figure 1

_____

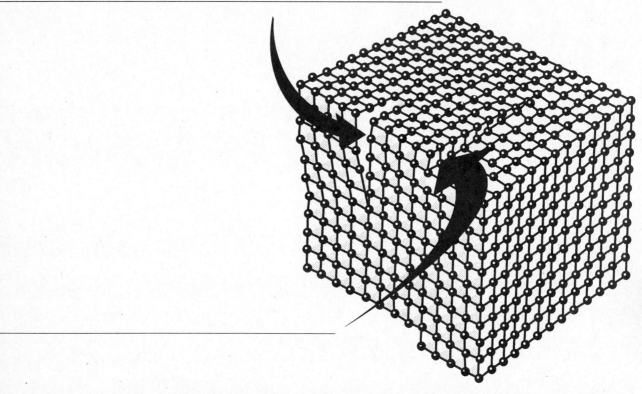

_____

Write T for true or F for false. If a statement is false, replace the underlined word or phrase with one that will make the statement true, and write your correction on the blank provided.

_____ 2. Imperfect crystals can be caused by defects to the unit cell, such as <u>misplaced atoms or ions</u>.

_____ 3. <u>Hydrated ions</u> are chemically bonded to water molecules.

_____ 4. $CuSO_4 \cdot 5H_2O$ and $CaSO_4 \cdot 2H_2O$ are <u>anhydrous</u> compounds.

_____ 5. It is <u>possible</u> to remove water molecules from hydrated compounds by raising the temperature or lowering the pressure.

_____ 6. Deliquescent materials are the <u>least</u> hygroscopic substances.

_____ 7. Materials that appear to be solid but that lack crystalline structures are called <u>amorphous</u> materials.

_____ 8. The symbol for amorphous materials is (<u>amph</u>).

_____ 9. Glass and molasses have <u>lower</u> viscosities than water and alcohol because they tend to flow less easily.

_____ 10. Metastable forms of an amorphous substance occur in <u>long-lasting</u> form.

Use the letters of the two-dimensional diagrams of (a) crystalline quartz and (b) quartz glass, shown in Figure 2, to answer the following questions.

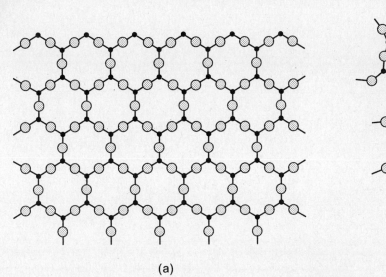
(a)

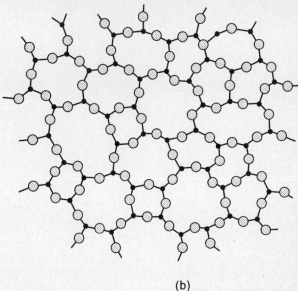
(b)

Figure 2

____ 11. Which substance shows the structure of an amorphous material?

____ 12. Which substance illustrates long-range order?

____ 13. Which substance has a specific melting point?

____ 14. Which substance can be classified as a supercooled liquid?

____ 15. Which substance is less stable in the form shown?

Chapter 17
STUDY GUIDE
17.1 CHANGES OF STATE

Complete the sentence or answer the question.

1. What is formed when molecules of a liquid or a solid substance escape from the surface of the substance? _____

2. According to kinetic theory, if the temperature of a liquid is lowered, the average velocity of its particles should _____.

3. When the ordered arrangement of the atoms of a solid substance breaks down, the solid _____.

4. When a liquid _____ the particles of the substance settle into an ordered arrangement and form a solid.

5. The equation $X(l) \rightarrow X(g)$ represents change from _____.

6. Write an equation that shows water, H_2O, as it changes reversibly between the liquid and the solid states. _____

7. According to Le Chatelier's principle, what happens if stress is applied to a system at equilibrium? _____

8. Write the name of the device shown in the figure.

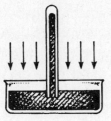

_____

9. What are the two devices shown in Figure 17.1 used to measure? _____

10. What is true about the vapor pressures of substances that have strong intermolecular forces? _____

11. What is meant by the melting point for a substance? _____

12. When dry ice is placed in an open container at room temperature, it changes directly from a solid to a vapor. This process is an example of _____.

13. The normal boiling point of a substance is the temperature at which its vapor pressure is equal to _____

14. A liquid that boils at a low temperature and evaporates rapidly at room temperature is described as being _____.

15. The temperature above which no amount of pressure will liquefy a gas is the _____ of the gas.

16. The condensation of substances that are normally gases is called _____.

17. Placing a checkmark in the correct column, identify whether each characteristic listed in the table applies to a substance with strong intermolecular forces or to a substance with weak intermolecular forces.

Intermolecular forces		
Characteristic	Strong	Weak
Volatile		
High boiling point		
High evaporation rate		
Low vapor pressure at room temperature		
Low critical temperature		

18. The specific heat of ice is 2.06 J/g·C°. Calculate how much energy is needed to heat a 52.0-g sample of ice from −20.0°C to 0.0°C.

19. The enthalpy of fusion of ice is 334 J/g. Calculate how much energy is needed to melt a 52.0-g sample of ice.

20. The C_p of liquid water is 4.18 J/g·C°. Calculate how much energy is needed to heat a 52.0-g sample of liquid water from 0.0°C to 100.0°C.

21. The enthalpy of vaporization of water is 2260 J/g. Calculate how much energy is needed to boil a 52.0-g sample of liquid water that is already at the boiling point.

22. The C_p of steam is 2.02 J/g·C°. Calculate how much energy is needed to heat 52.0-g of steam from 100.0°C to 120.0°C.

23. Use your calculations from questions 18 through 22 to calculate how much energy is necessary to convert a 52.0-g sample of ice at −20.0°C to steam at 120.0°C.

Chapter 17
STUDY GUIDE
17.2 SPECIAL PROPERTIES

Write T *for true and* F *for false. If a statement is false, replace the underlined word or phrase with one that will make the statement true, and write your correction on the blank provided.*

_____ 1. Water contracts when it freezes.

_____ 2. Compared with most molecules, water molecules have low masses.

_____ 3. Water is a gas at room temperature and standard atmospheric pressure.

_____ 4. Carbon dioxide and nitrogen are gases at room temperature and standard atmospheric pressure.

_____ 5. The structure of the molecules in a substance affects the interatomic and intermolecular forces that hold the substance together.

_____ 6. Molecules that contain hydrogen that is covalently bonded to a highly electronegative atom often have boiling points and melting points that are higher than would be expected in the absence of such hydrogen.

_____ 7. A molecule that is made up of a highly electronegative atom and a hydrogen atom is highly polar.

_____ 8. If an actual H^+ ion existed, it would consist of a bare neutron.

_____ 9. In compounds, hydrogen is always ionically bonded.

_____ 10. In a substance that is made up of identical polar molecules that contain a hydrogen atom, the hydrogen atom of each molecule in the substance is attracted to the positive portion of the other molecules.

_____ 11. A hydrogen bond results in a fairly strong hydrogen atom attachment between two molecules.

_____ 12. The attractive force between hydrogen-bonded molecules is much greater than the attractive force between other dipoles with the same electronegativity difference.

_____ 13. Hydrogen bonding is one type of dipole attraction.

_____ 14. Liquid water at 1°C is less dense than liquid water at 3°C.

_____ 15. Because ice is more dense than liquid water, ice floats at the surfaces of lakes and streams.

Match each term with the correct description.

a. hydrogen bonding
b. capillary rise
c. surface tension

___ 16. apparent elasticity at the surface of a liquid

___ 17. change in elevation of a liquid in a tube with a small diameter

___ 18. an attractive force that exists between molecules that contain hydrogen and a highly electronegative atom

19. A water molecule is a dipole. Look at the diagram of the water molecule. Place + and – signs in the blanks to label the areas of the molecule that are partially positive and partially negative.

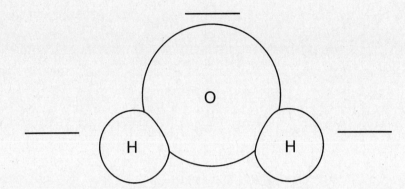

Chapter 18
STUDY GUIDE

18.1 VARIABLE CONDITIONS

Complete the sentence or answer the question.

1. According to the kinetic theory, a gas is made of very small particles that are in constant random _____.

2. What are the three factors on which the pressure exerted by a gas depends?
   _____
   _____

3. a. If the number of molecules in a constant volume increases, the pressure _____.
   b. If the number of molecules and the volume remain constant, but the kinetic energy of the molecules increases, the pressure _____.
   c. The kinetic energy depends on the _____.

4. Study the graph and answer the questions.
   a. What gas law is illustrated in the graph?
   _____
   b. Write a sentence describing the relationship represented by the graph.
   _____
   _____

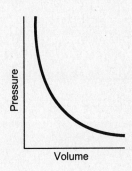

Write T for true or F for false. If a statement is false, replace the underlined word or phrase with one that will make the statement true, and write your correction on the blank provided.

_____ 5. When a gas is one in a mixture of gases, the pressure exerted by the individual gas is called its <u>ideal</u> pressure.

_____ 6. In calculations involving gases, the temperature given in Celsius must be converted to <u>Fahrenheit</u>.

_____ 7. At a constant pressure, the volume of a quantity of gas varies <u>directly</u> with the Kelvin temperature.

_____ 8. Gases collected by water displacement must be <u>soluble</u> in water.

_____ 9. <u>Charles's</u> law is represented by $P_{total} = P_1 + P_2 + \ldots P_n$.

Use Tables 18.1 and 18.2 in the text to answer questions 10 and 11.

10. What is the partial pressure of oxygen in the air at standard conditions?

11. a. What is the vapor pressure of water at 32°C? _____
    b. At what temperature is the vapor pressure of water 2.2 kPa? _____

12. Complete the table by using Boyle's law. In the column labeled *Change in volume,* indicate whether the volume increases or decreases. In the next column, write the two possible ratios for pressure and circle the appropriate ratio needed for calculations. Calculate the final volume.

Initial volume	Change in pressure	Change in volume	Pressure ratios	Final volume
26 cm³	55.8 kPa to 110.1 kPa			
131 dm³	225 kPa to 650.5 kPa			
88 dm³	36.8 kPa to 22.4 kPa			
925.0 cm³	151.2 kPa to 119.7 kPa			
0.621 m³	49.5 kPa to 76.2 kPa			

13. A 400-cm³ volume of gas is collected at 26°C. What volume would this gas occupy at standard conditions? Assume a constant pressure.

Match each equation below with the correct descriptions.

a. $P_{gas} = P_{total} - P_{water}$ b. $PV = k$ c. $P_{total} = P_1 + P_2 + \ldots P_n$ d. $V = kT$ e. $K = °C + 273.15$

___ 14. relationship between the Kelvin and Celsius temperature scales

___ 15. Boyle's law

___ 16. Charles's law

___ 17. Dalton's law

___ 18. used to find pressure of a gas collected over water

Chapter 18
STUDY GUIDE

18.2 ADDITIONAL CONSIDERATIONS OF GASES

Answer the questions in the spaces provided.

1. Why is it usually necessary to correct laboratory volumes of gas for both temperature and pressure?

2. How are the corrections made?

3. Why can the corrections for temperature and pressure be made in either order in one equation?

4. Why would people across the room from a newly opened bottle of perfume soon be able to smell the perfume?

5. Based on the diagrams in Figure 1, which gas diffuses faster, ammonia (top row) or carbon dioxide (bottom row)? _____

Figure 1

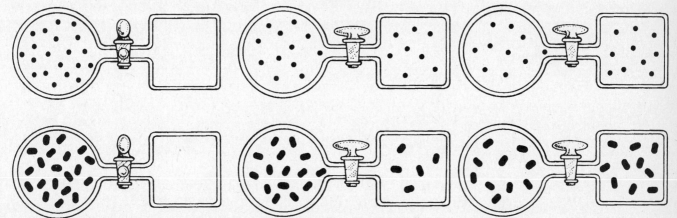

6. Calculate the relative rate of diffusion of the two gases.

7. a. Which gas diffuses faster, helium or neon?

   b. How much faster?

8. Using the illustrations for ammonia and carbon dioxide as guides, fill in the blank diagrams in Figure 2 comparing the rates of diffusion of helium and neon. Label your diagrams.

Figure 2

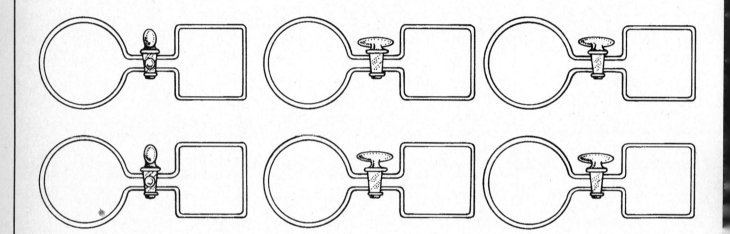

Chapter 19
STUDY GUIDE
19.1 Avogadro's Principle

Complete the sentence or answer the question.

1. What is Avogadro's principle?

2. Based on Avogadro's principle, if the volumes of two gases under similar conditions are equal, how are their number of moles related?

3. What is molar volume?

4. State what each of the five symbols in the ideal gas equation stands for.

5. What temperature scale is used to solve problems involving the ideal gas equation?

6. Using the ideal gas equation, complete the table by solving for the unknown variable. Show your work below.

P (kPa)	V (dm³)	n	T (K)
55.6	85.0	6.50	
	210.0	3.20	293
113	8.5		311
95.1		5.50	277

7. In the modified form of the ideal gas equation $PV = mRT/M$, what do the symbols m and M stand for?

   _____

8. How can this modified equation also be used to solve for the density of a gas?

   _____

9. Complete the table by solving for M. Show your work below.

P (kPa)	V (dm³)	m (g)	T (K)	M (g/mol)
98.6	6.27	0.832	300.0	
112.5	10.1	2.36	298	
88.7	23.5	0.55	292.5	
91.2	2.75	0.124	312	

Write T for true or F for false. If a statement is false, replace the underlined word or phrase with one that will make the statement true, and write your correction on the blank provided.

_____ 10. At a given <u>temperature</u>, the average kinetic energy of all gas molecules is the same.

_____ 11. One <u>gram</u> of oxygen will occupy 22.4 dm³ at STP.

_____ 12. The <u>Celsius</u> temperature scale is used for problems involving the ideal gas equation.

_____ 13. <u>Density</u> is equal to mass divided by volume.

_____ 14. The ideal gas equation combines <u>Boyle's</u> law and Charles's law.

Name _____ Date _____ Class _____

Chapter 19
STUDY GUIDE

19.2 GAS STOICHIOMETRY

Complete the sentence or answer the question.

1. What are the four steps you should keep in mind when solving mass-gas volume problems?
   _____
   _____
   _____

2. What are the four steps you should keep in mind when solving gas volume-mass problems?
   _____
   _____
   _____

Solve each of the following problems.

3. What volume of carbon dioxide, CO_2, at STP can be produced when 8.0 grams of oxygen, O_2, react with an excess of ethane, C_2H_6? Use the steps you listed in questions 1 and 2.

4. a. How many grams of carbon dioxide, CO_2, are formed if 96.0 g of oxygen, O_2, react with 12.2 dm^3 of ethylene, C_2H_4, to form carbon dioxide and water? Assume STP. Which reactant is the limiting one?

Chapter 19   STUDY GUIDE 75

b. How many grams of CO_2 are produced when 20.0 g of O_2 react with 31 dm^3 of ethylene? Assume STP. Which reactant is the limiting one?

Write T for true or F for false. If a statement is false, replace the underlined word or phrase with one that will make the statement true, and write your correction on the blank provided.

_____ 5. There is more than one way to find out which reactant is the <u>limiting reactant</u> in a chemical reaction.

_____ 6. <u>Volume-volume</u> relationships can be used when all the reactants and products are gases.

_____ 7. It is usually awkward to measure the <u>mass</u> of a gas.

_____ 8. A balanced equation tells you the <u>volume</u> ratios of the reactants and products.

_____ 9. In a chemical reaction, the amount of products is determined by the <u>excess</u> reactant.

Solve the following problems.

10. In a reaction between hydrogen and bromine to form hydrogen bromide gas, HBr, how many dm^3 of H_2 will be needed to completely use up 65.0 dm^3 of Br_2? (Assume constant pressure and temperature.)

11. How many moles and how many grams of HBr will be produced when the hydrogen and bromine react?

12. If only 22.4 dm^3 of H_2 reacted with 65 dm^3 of Br_2, how many grams of HBr would be produced?

Name _____ Date _____ Class _____

Chapter 20
STUDY GUIDE
20.1 SOLUTIONS

Write T for true or F for false. If a statement is false, replace the underlined word or phrase with one that will make the statement true, and write your correction on the blank provided.

_____ 1. Most solutions consist of a <u>gas</u> dissolved in a liquid.

_____ 2. <u>Molality</u> is the most common concentration unit in chemistry.

_____ 3. <u>Water</u> is the most common solvent.

_____ 4. An <u>amalgam</u> is a solid metal-metal solution.

_____ 5. The solubility of a substance can be changed by altering the <u>temperature</u>.

_____ 6. In terms of physical states of matter, there are <u>three</u> possible combinations of solvent-solute pairs.

_____ 7. Gases have <u>positive</u> enthalpies of solution.

_____ 8. The differing <u>solubilities</u> of substances can be used to separate them from mixtures.

_____ 9. Solvation <u>does not</u> occur in the case of a polar solvent and a nonpolar solute.

_____ 10. The solution process is usually <u>exothermic</u>.

Answer the questions below.

11. Why does a solution of potassium chloride not exhibit the characteristic behavior of potassium chloride?

_____

12. How is supersaturation possible?

_____
_____
_____

13. What are three things that you could do to speed up the rate of solution?

_____

14. What effect does pressure have on solutions in which the solute is a gas?

_____
_____

15. What is the difference between molarity and molality?

_____
_____

16. In terms of polarity, which solvent-solute combinations are most likely to form solutions?

_____

_____

17. Complete the table shown as a review of the solubility rules for water solutions. For each substance listed, place a check mark in the correct column indicating whether you would predict it to be soluble or insoluble in water. (Use Appendix Table A-7 for reference.)

Substance	Soluble	Insoluble
$Al_2(SO_4)_3$		
$PbBr_2$		
$ZnCO_3$		
$CaBr_2$		
$(NH_4)_3PO_4$		
$Mn(NO_3)_2$		
$SrSO_4$		
MgF_2		
$Co(CH_3COO)_3$		

18. Match each substance a-c with the reagent(s) that can be used to precipitate the positive ion. Write your answer(s) on the lines provided.

Reagents:

a. $SrBr_2$ _____

K_3PO_4

b. $MgCl_2$ _____

LiI

c. $AgNO_3$ _____

$Al_2(SO_4)_3$

Solve the following concentration problems.

19. If 10.0 g of sodium hydroxide is dissolved in 3.0 dm³ of water, what is the molarity of the NaOH solution formed?

20. What is the molality of a solution formed by dissolving 60.0 g of HNO_3 in 2.50×10^3 g of water?

21. If 150 g of *n*-butylacetate, $C_6H_{12}O_2$, is dissolved in 190 g of ethanol, C_2H_6O, what is the mole fraction of each component of the solution?

Name _____ Date _____ Class _____

Chapter 20
STUDY GUIDE
20.2 COLLOIDS

Match each type of colloid a through g with the correct example from the list that follows. Write the correct letter on the line provided. Some answers may be used more than once.

a. liquid emulsion
b. solid foam
c. solid sols
d. aerosols
e. solid emulsion
f. liquid foam
g. liquid sols

___ 1. mayonnaise

___ 2. whipped cream

___ 3. opals

___ 4. cheese

___ 5. jelly

___ 6. smoke

___ 7. pearls

___ 8. fog

___ 9. paint

___ 10. marshmallows

11. Complete the table comparing particles in solutions, suspensions, and colloids. Write *yes* or *no* for each characteristic to indicate whether it describes solutions, suspensions, or colloids.

Characteristic	Solutions	Suspensions	Colloids
less than 1 nm			
greater than 100 nm			
greater than 1 nm, but less than 100 nm			
settle out on standing			
pass unchanged through ordinary filter paper			
pass unchanged through membranes			
scatter light			
affect colligative properties			
particles can be seen with an ordinary light microscope			
heterogeneous			

Name _____ Date _____ Class _____

Answer each of the following questions on the lines provided.

12. What is colloid chemistry?
_____
_____

13. Why are colloids frequently used as the stationary phase in chromatography?
_____
_____

14. Why can semipermeable membranes be used to separate ions and colloidal particles?
_____

Match each property of colloids with the letter of the correct description. Write your answer on the line provided.

a. random motion of colloidal particles
b. scattering of a beam of light
c. attraction and holding of substances on the surfaces of dispersed particles
d. migration of positive and negative colloidal particles in an electric field

___ 15. Tyndall effect
___ 16. adsorption
___ 17. Brownian motion
___ 18. electrophoresis

19. The Figures 1 and 2 illustrate two characteristic properties of colloids. In each case, identify the property and explain why it occurs.

Figure 1

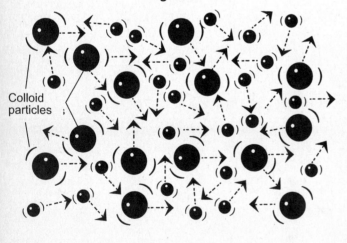

Figure 2

_____
_____
_____
_____

Chapter 21

STUDY GUIDE

21.1 VAPOR PRESSURE CHANGES

For each of the following, write T if the statement is true. If the statement is false, replace the underlined portion with a word or phrase that makes the statement true. Write your answer on the blank provided.

_____ 1. Vapor pressure, freezing point, and boiling point are <u>chemical</u> properties of a solution.

_____ 2. Colligative properties depend on the <u>kind</u> of particles present in solution.

_____ 3. If fewer solvent particles can evaporate from a solution, the vapor pressure of the solvent is <u>lowered</u>.

_____ 4. For a solution having a specific concentration, the amount that the vapor pressure of the solvent is lowered depends on the characteristics of the <u>solute</u>.

_____ 5. Glucose, sucrose, and other molecular compounds with high melting points are considered <u>volatile</u> solutes.

_____ 6. The amount the vapor pressure of a solvent is lowered is equal to the difference between the vapor pressure of the pure solvent and the vapor pressure of the <u>solution</u>.

_____ 7. According to Raoult's law, the vapor pressure of a solution is directly proportional to the mole fraction of <u>solute</u>.

_____ 8. The vapor pressure of a sugar solution in which the mole fraction of sugar is 0.25 will be <u>4</u> times the vapor pressure of pure water.

_____ 9. In an ideal solution, all possible attractions among particles of solute and solvent are <u>the same</u>.

_____ 10. In an aqueous sucrose solution at 0°C, the mole fraction of water is 0.85. If the vapor pressure of water at 0°C is 0.6 kPa, its vapor pressure is lowered by <u>0.09</u> kPa.

Use the diagram of a fractional distillation apparatus shown here to answer the following questions about a solution consisting of pure liquids X and Y, where the boiling point of liquid X is greater than that of liquid Y.

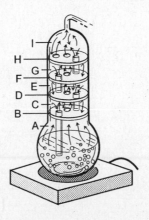

11. Which of the substances in the given solution is more volatile? Why?

_____

12. When the vapor at A is in equilibrium with the given solution, is the vapor richer in substance X or substance Y?

13. The substances at D and E are in equilibrium. How does the temperature of the solution at D compare with the temperature of the vapor at E?

14. How does the temperature at I compare with the temperature at A? Explain your reasoning.

15. How does the composition of the substances at the top of the apparatus (H and I) differ from the composition of the substances at B and C?

Use Table 21.1 in your textbook to answer the following questions about the fractional distillation of petroleum.

16. Which fraction(s) would come off first? Explain your reasoning.

17. Which fraction(s) would collect at the bottom of the fractionating tower?

18. Would the fractions at the middle levels of the tower be richer in fuel oil or jet fuel?

Use the symbol > (greater than) or < (less than) to complete each of the following comparisons between vapor pressures, boiling points, and freezing points of (1) a pure solvent, and (2) a solution of a nonvolatile solute in the solvent at the same temperature.

19. vapor pressure: pure solvent ___ solution
20. boiling point: pure solvent ___ solution
21. freezing point: pure solvent ___ solution

Answer each of the following questions.

22. Describe and explain the effect of adding a nonvolatile, nonionizing solute on the boiling point of a pure solvent.

23. Why is the addition of antifreeze to the water in a car radiator just as important in hot climates as in cold climates?

Name _____ Date _____ Class _____

Chapter 21
STUDY GUIDE

21.2 QUANTITATIVE CHANGES

Answer each of the following questions.

1. What is meant by the molal boiling point constant and molal freezing point constant for a given solvent?
   _____
   _____
   _____

2. Describe freezing point depression in terms of the freezing points of a pure solvent and a solution made from that solvent.
   _____
   _____

3. Ideally, a $1m$ solution of zinc chloride ($ZnCl_2$) in water would boil at 101.545°C, whereas a $1m$ solution of glucose ($C_6H_{12}O_6$) would boil at 100.515°C.

   a. What is the boiling point elevation of an ideal solution of zinc chloride in water?
   _____

   b. Compare the boiling point elevation of a zinc chloride solution with the boiling point elevation of a glucose solution.
   _____

   c. How does the difference in boiling point elevation provide evidence that zinc chloride dissociates in water?
   _____
   _____
   _____

4. The molal freezing point constant for pure water is 1.853 C°/m. Assuming 100% dissociation, at what temperature would a $1m$ solution of magnesium chloride ($MgCl_2$) in water freeze?
   _____

5. A solution contains 20.0 g maltose ($C_{12}H_{22}O_{11}$), a nonvolatile, nonionizing solute, dissolved in 324 g H_2O. Fill in the missing parts and complete the following calculations for the boiling and freezing points of the resulting solution.

   a. $\dfrac{20.0 \text{ g } C_{12}H_{22}O_{11}}{324 \text{ g } H_2O} \Big| \dfrac{1 \text{ mol } C_{12}H_{22}O_{11}}{_____} \Big| \dfrac{_____}{1 \text{ kg } H_2O} = $ _____

   b. boiling point elevation = (_____)(0.515 C°/m) = _____
      boiling point = _____ °C

   c. freezing point depression = (_____) (1.853 C°/m) _____
      freezing point = _____ °C

Name _____ Date _____ Class _____

In the diagrams shown, a semipermeable membrane covers the opening of the test tube that is inverted in the beaker. The molecules of solvent used can pass through the membrane, but the molecules of solute cannot. Study the diagrams, then answer the questions that follow.

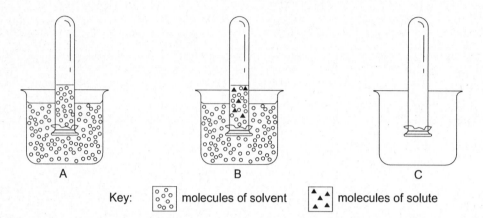

Key: molecules of solvent   molecules of solute

6. Draw arrows on diagram A to indicate the direction and rate of movement of solvent molecules. Explain the reasoning for what you drew.
   _____

7. Describe the direction and rate of movement of solvent and solute molecules in diagram B. Draw arrows to represent the movement of solvent molecules.
   _____
   _____

8. In terms of solvent concentration, describe the net movement of molecules in diagram B.
   _____

9. On which side of the membrane in diagram B will an increase in volume take place?
   _____

10. At equilibrium, on which side of the membrane will osmotic pressure be exerted?
    _____

11. On diagram C, show the solution and the pure solvent at equilibrium. Use arrows to indicate the flow of solvent molecules.

For each of the following, write T if the statement is true. If the statement is false, replace the underlined portion with a word or phrase that makes the statement true. Write your answer on the line provided.

_____ 12. Ion activity is an indication of the degree to which an ionic compound <u>actually</u> dissociates in a water solution.

_____ 13. In reality, ionizing ionic compounds in aqueous solution do not completely dissociate because of the <u>repulsive</u> forces between <u>like</u> charged ions.

_____ 14. For an ionizing ionic compound dissolved in water, the more dilute the solution, the <u>less</u> effective each ion is in freezing point depression and boiling point elevation.

84 STUDY GUIDE   Chapter 21

Chapter 22
STUDY GUIDE

22.1 REACTION RATES

Write T for true or F for false. If the statement is false, replace the underlined word or phrase with one that makes the statement true and write your correction on the blank provided.

_____ 1. A substance that does not break down spontaneously at room temperature is said to be <u>kinetically unstable</u>.

_____ 2. A reaction that eventually reaches equilibrium is said to be <u>reversible</u>.

_____ 3. A reaction goes to completion if at least one of the <u>reactants</u> is used up.

_____ 4. The rate of a reaction from left to right is the rate at which a product <u>disappears</u>.

_____ 5. Heat, a flame, or a spark may supply the <u>activated complexes</u> needed for a reaction to take place.

_____ 6. The reactions of kinetically stable substances tend to have high <u>activation energies</u>.

_____ 7. The energy required to form an activated complex is converted to <u>kinetic</u> energy.

_____ 8. Ionic reactions tend to have <u>faster</u> reaction rates than electron transfer reactions.

_____ 9. Reaction rate varies directly as the product of the <u>masses</u> of the reactants.

_____ 10. Specific rate constant refers to the rate at which a reaction occurs at a given <u>temperature</u>.

_____ 11. For a given reaction, the total number of molecules having the required activation energy at 125°C would be <u>the same as</u> the total number of molecules having the required activation energy at 100°C.

_____ 12. A substance that affects reaction rate without being chemically changed at the end of the reaction is <u>an inhibitor</u>.

Match each situation in questions 13-18 with the letter of the term that best describes it. A letter may be used more than once.

a. homogeneous catalyst
b. homogeneous reaction
c. heterogeneous catalyst
d. heterogeneous reaction
e. inhibitor
f. kinetically stable
g. thermodynamically stable

____ 13. $H_2O(g) + CO(g) \rightarrow H_2(g) + CO_2(g)$

____ 14. The reaction between $N_2(g)$ and $O_2(g)$ does not take place at room temperature because high activation energies are required.

____ 15. $CaO(cr) + CO_2(g) \rightarrow CaCO_3(cr)$

____ 16. Manganese dioxide powder is added to increase the decomposition rate of liquid potassium chlorate.

____ 17. BHT is added to food as a preservative.

____ 18. $C_2H_5OH(aq) + CH_3COOH(aq) \rightleftharpoons CH_3COOC_2H_5(aq) + H_2O(l)$

Name _____ Date _____ Class _____

For each of the following reactions, write I *if the effect of the given change is an increase in reaction rate,* D *if the effect is a decrease in reaction rate, or* RS *if the reaction rate remains the same. Assume that the reaction takes place in a closed vessel of fixed volume.*

19. Limestone (calcium carbonate) reacts with hydrochloric acid in an irreversible reaction, to form carbon dioxide and water as described by the following equation:
$CaCO_3(cr) + 2HCl(aq) \rightarrow CaCl_2(aq) + CO_2(g) + H_2O(l)$
What is the effect if

_____ a. the temperature is lowered?

_____ b. the volume of the reaction vessel is increased?

_____ c. limestone chips are used instead of a block of limestone?

_____ d. the pressure inside the reaction vessel is increased?

_____ e. a more dilute solution of HCl is used?

20. At a given temperature, ethene, C_2H_4, reacts with chlorine in a reversible reaction, to produce 1,2-dichloroethane ($C_2H_4Cl_2$) as described by the following equation:
$C_2H_4(g) + Cl_2(g) \rightarrow C_2H_4Cl_2(g)$
What is the effect if

_____ a. the volume of the reaction vessel is reduced?

_____ b. the pressure inside the reaction vessel is reduced?

_____ c. a catalyst is added?

_____ d. the volume of the reaction vessel is doubled and the pressure inside the vessel is halved?

Use the energy diagram shown to answer questions 21 - 29.

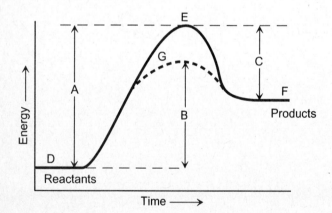

_____ 21. Which letter indicates the energy content of the reactants?

_____ 22. Which letter indicates the energy content of the products?

_____ 23. Which letter indicates the change in energy required for the formation of activated complexes in the forward reaction?

_____ 24. Assume that the reaction is reversible. Which line indicates the change in energy required for the formation of activated complexes in the reverse reaction?

_____ 25. Which letter represents the energy of the activated complexes in the uncatalyzed reaction?

_____ 26. Which letter indicates the reaction in the presence of a catalyst?

_____ 27. Which letter indicates the activation energy needed for the forward reaction if a catalyst is used?

28. How are the activation energies of the forward and reverse reactions affected by the addition of a catalyst?

29. How does the potential energy of the activated complexes of the forward reaction compare with the potential energy of the activated complexes of the reverse reaction?

30. For the reaction A + B → C, the following data were obtained:

Trial	[A]	[B]	Rate
1	0.10M	0.20M	0.00003 (mol/dm^3)/s
2	0.10M	0.40M	0.00006 (mol/dm^3)/s
3	0.20M	0.40M	0.00048 (mol/dm^3)/s

a. How is the rate of reaction affected by a change in concentration of reactant A? Reactant B?

b. If the rate expression for this reaction is: rate = $k[A]^3[B]$, find k.

Chapter 22
STUDY GUIDE
22.2 CHEMICAL EQUILIBRIUM

For items 1-8, underline the term inside the parentheses that makes each statement true.

1. At equilibrium, the rate of the forward reaction is (equal to, greater than) the rate of the reverse reaction.

2. The equilibrium constant for a given reaction at a given temperature is the (product, quotient) of the specific rate constant for the forward reaction and the specific rate constant for the reverse reaction.

3. The exponents used in the expression for the equilibrium constant are the (subscripts, coefficients) of the reactants and products.

4. If a reaction tends to go toward completion, the rate of the forward reaction is (equal to, greater than) the rate of the reverse reaction before equilibrium is reached.

5. If $K_{eq} = 1.2 \times 10^{-5}$, the concentration of the reactants is (greater than, less than) the concentration of the products at equilibrium.

6. At temperature T_1, K_{eq} for a certain reaction is 0.239. At temperature T_2, K_{eq} for the same reaction is 4.7. By changing the temperature from T_1 to T_2, the equilibrium will shift in favor of the (reactants, products).

7. If the reaction $H_2(g) + Cl_2(g) \rightleftarrows 2HCl(g) + heat$ is at equilibrium, a decrease in (volume, temperature) will produce a shift in equilibrium toward the right.

8. An increase in pressure on the system $2CO_2(g) \rightleftarrows 2CO(g) + O_2(g)$ at equilibrium results in an equilibrium shift toward the (left, right).

Answer or complete each of the following items.

9. What factors can affect the equilibrium of a reaction?

_____

10. A reversible one-step reaction occurs between carbon monoxide gas, CO, and hydrogen gas, H_2, to produce methane gas, CH_4, and gaseous water. Using this information, fill in the diagram according to the following guidelines:
    a. Within the ovals, write the balanced equation for the reaction.
    b. Label the forward and reverse reactions on the lines provided.
    c. In the rectangles, write the words *reactants* or *products*, as appropriate, to represent both the forward and reverse reactions.

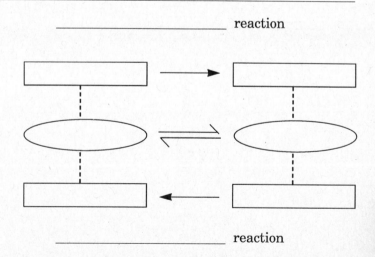

11. Write the expression for the equilibrium constant for the reaction in question 10.

12. If temperature and pressure of the reaction in question 10 are kept constant, but the concentration of each of the substances is halved,
    a. how does the rate of the forward reaction change?

    b. how does the rate of the reverse reaction change?

    c. what would be the net change in the relative amounts of reactants and products?

13. In the reaction in question 10, if the volume of the reaction vessel and temperature are kept constant and the pressure on the system is increased,
    a. the concentrations of which substances would be affected? _____
    b. in which direction will the equilibrium shift? _____
    c. which substance(s) will show an increase in concentration when equilibrium is reestablished?

14. Consider the equilibrium equation for the reaction: $4HCl(g) + O_2(g) + heat \rightleftarrows 2Cl_2(g) + 2H_2O(g)$
    a. If the temperature is increased, the reaction will favor the formation of
    _____.
    b. Which reaction requires an input of energy? _____

15. In the Haber process, which involves the reaction $N_2(g) + 3H_2(g) \rightleftarrows 2NH_3 + energy$,
    a. why is NH_3 removed as it is formed?

    b. why is the use of a catalyst considered one of the optimum conditions for this process?

    c. what is the effect on the relative amounts of product and reactant if the catalyst is removed?

16. Given $K_{eq} = \dfrac{[NO]^4[H_2O]^6}{[NH_3]^4[O_2]^5}$

    a. Write the chemical equation for the reversible reaction having the given K_{eq}.

    b. At a certain temperature the concentrations of NO and NH_3 are equal, and the concentration of H_2O and O_2 are 2.0M and 3.0M respectively. What is the value of K_{eq} at this temperature?

Name _____ Date _____ Class _____

Chapter 23
STUDY GUIDE

23.1 ACIDS AND BASES

Complete the sentence or answer the question.

1. Define or explain the following theories about acids and bases.

   a. Arrhenius Theory _____
   _____
   _____

   b. Brønsted-Lowry Theory _____
   _____

2. A(n) _____ is an acid, base, or salt, which, when dissolved in water, conducts an electric current.

3. The polyatomic ion, H_3O^+, which is formed when a hydrogen ion (H^+) combines with a water molecule, is called the _____.

4. The _____ of an acid is the particle that remains after a proton has been released by the acid.

5. The _____ of a base is formed when the base acquires a proton from the acid.

6. Complete the following equation that, *in general*, describes any acid-base reaction:

   _____ + _____ → _____ + _____

7. Acids that contain only two elements are called _____.

8. Acids that contain three elements are called _____.

9. Substances that can react as either an acid or a base are said to be _____.

10. The most common example of the kind of compound described in exercise 9 is _____.

Match each term with the correct description.

a. acidic anhydride   d. strong acid
b. basic anhydride   e. strong base
c. weak acid   f. weak base

___ 11. a base that is completely dissociated in solution into positive ions and negative ions

___ 12. any oxide that will produce an acid when dissolved in water

___ 13. an acid that is considered to ionize completely in solution into positive ions and negative ions

___ 14. any oxide that will produce a base when dissolved in water

___ 15. slightly ionized acid in solution

___ 16. a base that is only partially ionized in solution

Name ________________ Date ________________ Class ________________

Complete the following exercises.

17. Name the following binary acids.

a. HF ________________

b. HCl ________________

c. HI ________________

d. H_2S ________________

e. HN_3 ________________

f. HBr ________________

18. Name the following ternary acids and bases.

a. H_3AsO_4 ________________

b. H_3AsO_3 ________________

c. NaOH ________________

d. H_3BO_3 ________________

e. H_3PO_4 ________________

f. H_3PO_3 ________________

g. HNO_3 ________________

19. Name the following organic acids and bases.

a. CH_3NH_2 ________________

b. CH_3COOH ________________

c. CH_3CH_2COOH ________________

20. Write the formula for the conjugate base of each of the acids listed.

a. H_2O ________________

b. HNO_3 ________________

c. HF ________________

d. $HC_2H_3O_2$ ________________

21. Write the formula for the conjugate acid of each of the following bases listed.

a. NH_3 ________________

b. HSO_4^- ________________

c. HS^- ________________

d. C_2H_5NH ________________

Name ______________________ Date ______________ Class ______________

22. Classify each of the following as a strong acid, strong base, weak acid, or weak base.

a. NaOH ______________

b. HCl ______________

c. NH_4^+ ______________

d. NH_3 ______________

e. Cl^- ______________

f. HI ______________

23. In the following reaction, which species behave as Brønsted acids? As Brønsted bases?
$H_2SO_4(aq) + H_2O \rightleftarrows HSO_4^-(aq) + H_3O^+(aq)$

____

____

24. Write the formulas for the anhydrides of the following.

a. H_2CO_3 ______________

b. H_2SO_3 ______________

c. H_2SO_4 ______________

d. NaOH ______________

e. $Ca(OH)_2$ ______________

f. HNO_3 ______________

Name _____ Date _____ Class _____

Chapter 23
STUDY GUIDE

23.2 SALTS AND SOLUTIONS

Complete the sentence or answer the question.

1. Define or explain the following terms.

    a. salt _____

    b. neutralization reaction _____

    c. polyprotic acid _____

2. Ions present in a solution but not involved in the reaction are called _____.

3. State the general rule that must be observed when writing net ionic equations.
   _____
   _____

4. Convert the following balanced equations first to ionic form and then to net ionic form.

    a. $BaCl_2 + Na_2SO_4 \rightarrow BaSO_4 + 2NaCl$

    b. $Na_2CO_3 + 2HCl \rightarrow 2NaCl + CO_2 + H_2O$

    c. $NiS + 2HCl \rightarrow NiCl_2 + H_2S$

5. Select the polyprotic acids from the following list.

    HCN, H_2CO_3, HCl, HF, H_2SO_3, H_3PO_4, H_2SeO_3, HBr
    _____

Write T for true or F for false. If a statement is false, replace the underlined word or phrase with one that will make the statement true, and write the correction on the blank provided.

_____ 6. A salt is formed as a product in the <u>equilibrium</u> reaction of an acid and a base.

_____ 7. <u>Salts</u> may be formed from the reaction of acidic or basic anhydrides with a corresponding acid, base, or anhydride.

_____ 8. The percent ionization of a weak acid or base dissolved in water is a measure of the amount of <u>ionization</u> of the acid or base in solution.

Answer each question.

9. Write the equilibrium constant expression, K_{eq}, for the following acid ionization reaction.
   $HOBr + H_2O \rightleftarrows H_3O^+ + OBr^-$

10. Write the ionization constant expression, K_a, for the reaction in question 9.

11. Calculate the hydronium ion concentration of a 0.25M solution of hypochlorous acid, HOCl, for which $K_a = 3.5 \times 10^{-8}$. The equation for ionization of this acid is $HOCl + H_2O \rightleftarrows H_3O^+ + OCl^-$.

12. Calculate the percent of ionization of 0.010M acetic acid solution, if the hydronium ion concentration is $4.2 \times 10^{-4} M$.

Chapter 24 STUDY GUIDE

24.1 WATER EQUILIBRIA

Complete the sentence or answer the question.

1. Define or explain the following. (Include formulation where applicable.)

    a. solubility product constant, K_{sp} of a saturated solution _____

    b. ion product constant of water, K_w _____

    c. pH _____

    d. hydrolysis _____

    e. buffer solution _____

2. The addition of a common ion to a saturated solution _____ the solubility of a substance in solution.

3. Write the solubility product expression for each of the following.

    a. $BaSO_4$ _____ d. $MgNH_4PO_4$ _____
    b. CaF_2 _____ e. Ag_2CrO_4 _____
    c. Bi_2S_3 _____

4. The concentration of water in pure water is _____ mol/dm³.

5. Write the formula for calculating the pH from the concentration of H_3O^+.

6. Find the pH of solutions with the following H_3O^+ concentrations.
   a. $4.21 \times 10^{-4} M$

   b. $5.00 \times 10^{-6} M$

   c. $9.11 \times 10^{-8} M$

   d. $2.14 \times 10^{-13} M$

   e. $5.21 \times 10^{-10} M$

   f. $1.45 \times 10^{-2} M$

Match the acid-base reaction type with the examples given.

a. weak base + strong acid
b. strong acid + strong base
c. weak acid + strong base

___ 7. $H^+ + OH^- \rightleftarrows H_2O$

___ 8. $HC_2H_3O_2 + OH^- \rightleftarrows H_2O + C_2H_3O_2^-$

___ 9. $NH_3 + H^+ \rightleftarrows NH_4^+$

10. Using the rules outlined in Chapter 24, classify each of the following solutions as acidic, basic, or neutral.
    a. $NaNO_3$ _____   d. NH_4Br _____
    b. KCN _____   e. $AlCl_3$ _____
    c. Na_2CO_3 _____

Write T for true or F for false. If a statement is false, replace the underlined word or phrase with one that will make the statement true, and write your correction on the blank provided.

_____ 11. Buffers are <u>most</u> efficient at neutralizing added acids or bases when the concentrations of HA and A⁻ (or MOH and M⁺) are the same.

_____ 12. The pH is the measure of <u>oxygen</u> in a solution.

_____ 13. Pure water is a <u>good</u> conductor of electricity.

_____ 14. $\underline{K_w}$ is a constant for all dilute aqueous solutions at room temperature.

_____ 15. As the hydronium concentration increases in a solution, the solution is made more <u>basic</u>.

Answer each question.

16. Label the parts of the pH scale in terms of acid, base, and neutral segments.

    pH = 0 ◄─────────────── pH = 7 ───────────────► pH = 14

17. What is the concentration of silver in a saturated solution of $AgC_2H_3O_2$? The K_{sp} of $AgC_2H_3O_2$ is 2.00×10^{-3}.

18. Will a precipitate of $CaSO_4$ form in **hard water** if the Ca^{2+} concentration is $0.010M$ and the SO_4^{2-} concentration is $0.00100M$? Show your calculations. (K_{sp} of $CaSO_4 = 3.0 \times 10^{-5}$)

19. If 0.40 dm³ of $0.020M$ Na_2SO_4 is added to 0.80 dm³ of hard water, where $[Ca^{2+}] = 0.010M$ (as in question 18), will a precipitate form? Show your calculations.

Name _____ Date _____ Class _____

Chapter 24
STUDY GUIDE

24.2 TITRATION

Complete the sentence or answer the question.

1. Define or explain the following.

   a. indicator _____
   _____

   b. titration _____
   _____

   c. standard solution _____

2. Explain what a pH meter is and what it is used for.
   _____
   _____

3. Explain how a titration is carried out.
   _____
   _____
   _____
   _____

4. A titration is performed in which 22.3 cm³ of 0.240M NaOH reacts with a 50.0-cm³ sample of $HC_2H_3O_2$. What is the concentration of the $HC_2H_3O_2$?

Write T for true or F for false. If a statement is false, replace the underlined word or phrase with one that will make the statement true, and write your correction on the blank provided.

_____ 5. Titration indicators are most useful when they are used on <u>colorful</u> solutions.

_____ 6. If an acid is added to a base, a <u>neutralization reaction</u> occurs.

_____ 7. Many indicators must be used in order to test for pH over a <u>wide</u> range of the pH scale.

100   STUDY GUIDE

Chapter 25
STUDY GUIDE

25.1 OXIDATION AND REDUCTION PROCESSES

Complete each of the following.

1. A reaction in which ions or atoms undergo changes in electron structure is referred to as a(n) _____ reaction.
2. The loss of electrons from an atom or ion is called _____.
3. The gain of electrons by an atom or ion is called _____.
4. The substance in a redox reaction that undergoes oxidation is referred to as a(n) _____ agent.
5. The substance in a redox reaction that undergoes reduction is referred to as a(n) _____ agent.
6. In the reaction that follows, _____ is oxidized and _____ is reduced. $H_2 + Cl_2 \rightarrow 2HCl$
7. The oxidation number of a free element is _____.
8. In most reactions, the oxidation number of the calcium ion is _____, while that for the oxide ion is generally _____.
9. The sum of the oxidation numbers of all of the atoms in HNO_3 is _____.
10. The oxidation number of N in $Mg(NO_3)_2$ is _____.
11. The oxidation number of S in SCl_2 is _____.
12. In the reaction that follows, the oxidizing agent is _____. $C + H_2O \rightarrow CO + H_2$
13. In the reaction given in question 12, the element being oxidized is _____.
14. In the reaction given in question 12, the total number of electrons transferred (lost or gained) is _____.
15. The oxidation number of P in PO_4^{3-} is _____.
16. Write true (T) or false (F) for each of the following.

    ___ a. The following is an example of a redox reaction: $HCl + NaOH \rightarrow NaCl + H_2O$

    ___ b. The term *oxidation* was first applied to the combining of oxygen with other elements.

    ___ c. In an oxidation-reduction reaction, the number of electrons lost must equal the number gained.

    ___ d. If a substance gains electrons readily, it is said to be a strong reducing agent.

    ___ e. The oxidation number of a monatomic ion is equal to the charge on the ion.

    ___ f. In general, ions of the Group 17 elements have oxidation numbers of 1–.

    ___ g. The oxidizing agent in the following reaction is Na. $2Na + S \rightarrow Na_2S$

    ___ h. A total of 2 electrons are transferred in the reaction given in question 16g.

    ___ i. The element that undergoes reduction in the reaction given in question 16g is S.

17. Complete the table below for each reaction specified.

Reaction	Element oxidized	Element reduced	Oxidizing agent	Reducing agent	Total e^- transferred
$2HBr + Cl_2 \rightarrow$ $2HCl + Br_2$					
$Zn + I_2 \rightarrow ZnI_2$					
$Fe_2O_3 + 3CO \rightarrow$ $2Fe + 3CO_2$					
$16H^+ + 2MnO_4^- + 5C_2O_4^{2-} \rightarrow$ $2Mn^{2+} + 8H_2O + 10CO_2$					

Chapter 25
STUDY GUIDE

25.2 BALANCING REDOX EQUATIONS

1. Complete the steps outlined below in order to balance the following oxidation-reduction reaction:
   $HNO_3 + S \rightarrow NO_2 + H_2SO_4 + H_2O$

   a. Write the skeleton equation for the reaction.

   b. Assign oxidation numbers to all atoms involved in the reaction.

   c. Identify the substance being oxidized, and then write and balance the oxidation reaction in terms of both atoms and charge. In acidic solutions, it may be necessary to add H^+ ions and H_2O molecules to balance the reaction. In basic solutions, it may be necessary to add a sufficient number of OH^- ions to each side of the equation to combine with any excess H^+ ions and form H_2O molecules.

   d. Identify the substance being reduced, and then write and balance the reduction reaction in terms of both atoms and charge.

   e. Combine the two half-reactions, first multiplying one or both by the factor(s) needed to balance the electron transfer. Cancel the electrons, as well as the excess number of any species that appear on both sides of the equation, from the final reaction.

   f. Perform a final check on the balanced equation to ensure that both atoms and charge are balanced.

Apply the steps given above to balance each of the following oxidation-reduction reactions.

2. $Br_2 + SO_2 + H_2O \rightarrow H_2SO_4 + HBr$

3. $PbS + H_2O_2 \rightarrow PbSO_4 + H_2O$

4. $H_3AsO_4 + Zn \rightarrow AsH_3 + Zn^{2+}$

5. $PH_3 + O_2 \rightarrow P_4O_{10} + H_2O$

6. $NO_2 + OH^- \rightarrow NO_2^- + NO_3^-$ (in basic solution)

7. $K_2Cr_2O_7 + Na_2SO_3 + HCl \rightarrow Cr_2(SO_4)_3 + KCl + NaCl + H_2O$

8. $Cu(OH)_2 + HPO_3^{2-} \rightarrow Cu_2O + PO_4^{3-}$

9. $TeO_2 + BrO_3^- \rightarrow H_6TeO_6 + Br_2$

10. $Bi_2S_3 + NO_3^- \rightarrow Bi^{3+} + NO + S$

11. $NO + H_5IO_6 \rightarrow NO_3^- + IO_3^-$

Chapter 26
STUDY GUIDE

26.1 CELLS

1. Indicate if the bulb shown in Figure 1 will light or not by placing a plus(+) or minus(−) sign, respectively, in the blank to the left of each of the following situations.

   ___ a. The electrodes are connected with a piece of glass.
   ___ b. The beaker is filled with air.
   ___ c. The beaker is filled with an electrolyte.
   ___ d. The electrodes are connected by a wire.
   ___ e. The beaker is filled with alcohol.

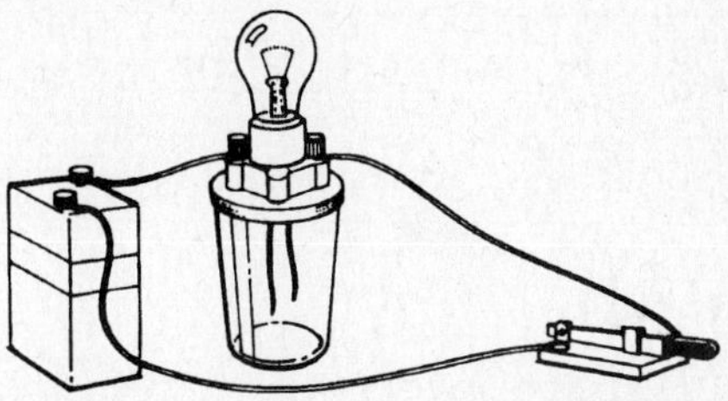

   Figure 1

2. In situation c of question 1, write the type of conduction, if any, of the materials.

   a. wires _____   c. bulb filament _____
   b. electrolyte _____   d. electrodes _____

Write T for true or F for false. If a statement is false, replace the underlined word or phrase with one that will make the statement true, and write your correction on the blank provided.

_____ 3. A galvanometer is an instrument that detects <u>electric current</u>.

_____ 4. Potential difference is measured in units of <u>amperes</u>.

_____ 5. Current flows through wires in which there is <u>no</u> potential difference between the ends.

_____ 6. Metallic conduction takes place because the electrons of metals <u>are free to move</u> when a small potential difference is applied.

_____ 7. Electrolytic conduction takes place because <u>electrons</u> move freely in aqueous solutions.

Complete the sentence.

8. An _____ is any substance that produces ions in solution.

9. The process by which an electric current produces a chemical change is called _____.

10. In a cell, oxidation and reduction occur as separate half-reactions at the same _____.

11. A device that converts chemical potential energy into electric energy is called a _____.

12. The zinc-copper cell reaction, in which Zn is oxidized to Zn^{2+} and Cu^{2+} is reduced to Cu, is represented as _____.

Using the letters of the terms below, label each of the diagrams in Figure 2.

a. anode (Pb)
b. anode (Zn can)
c. anode (Zn shell)
d. automobile battery
e. cathode (graphite)
f. cathode (PbO$_2$)
g. cathode (steel)
h. dry cell
i. electrolyte (H$_2$SO$_4$)
j. electrolyte (KOH and paste of Zn(OH)$_2$ and HgO)
k. electrolyte (paste of MnO$_2$, NH$_4$Cl, and powdered graphite)
l. mercury battery

Figure 2

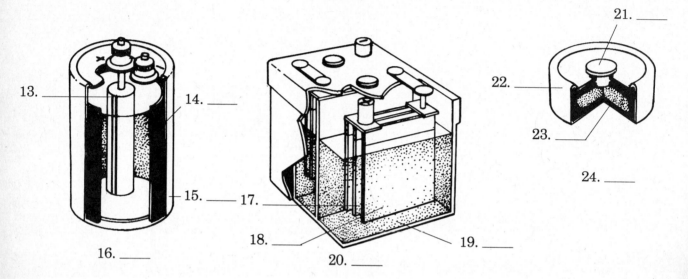

Chapter 26
STUDY GUIDE
26.2 QUANTITATIVE ELECTROCHEMISTRY

Write T for true or F for false. If a statement is false, replace the underlined word or phrase with one that will make the statement true, and write your correction on the blank provided.

_____ 1. The difference between two half-cells is a measure of the relative tendency of the two substances to take on <u>protons</u>.

_____ 2. The half-reactions listed in Table 26.1 of the text are written as <u>oxidations</u>.

_____ 3. The symbol for standard reduction potential (25°C, 101.325 kPa, 1M) is <u>E^0</u>.

_____ 4. The half-reaction of <u>carbon</u> is assigned a reduction potential of 0.000 0 V.

_____ 5. The standard reduction potential of a substance is an <u>intensive</u> property.

Use Table 26.1 in the text to answer questions 6-11.

6. Rank the following substances in order of decreasing reduction strength.
   Ag^+ Au^+ Cs^+ K^+ Li^+ Na^+

   _____

7. Explain why the reaction, $Cu(cr) + 2Ag^+(aq) \rightarrow 2Ag(cr) + Cu^{2+}(aq)$ will occur spontaneously.

   _____
   _____
   _____

8. Predict if each of the following reactions will occur spontaneously as written.

   a. $Cu(cr) + Pd^{2+}(aq) \rightarrow Cu^{2+}(aq) + Pd(cr)$

   b. $2Ag(cr) + Co^{2+}(aq) \rightarrow Co(cr) + 2Ag^+(aq)$

9. Consider the following cell (standard conditions).
   $Zn\,|\,Zn^{2+}\,||\,Ni^{2+}\,|\,Ni$

   a. What substance is oxidized in the cell?

   b. Write the oxidation half-cell reaction.

   c. What is the potential for the oxidation half-cell reaction?

   d. What substance is reduced in the cell?

   e. What is the reduction half-cell reaction?

   f. What is the potential for the reduction half-cell reaction?

   g. What is the potential produced by the cell?

10. Predict the potential for each of the following cells.
    a. $Zn\,|\,Zn^{2+}\,||\,Cu^{2+}\,|\,Cu$

    b. $Fe\,|\,Fe^{2+}\,||\,Ag^{+}\,|\,Ag$

11. The relationship $E = -0.05916\,pH$ indicates that E is a _____ function of pH.

12. The following cell has a standard potential of 0.459V.
    $Cu\,|\,Cu^{2+}\,||\,Ag^{+}\,|\,Ag$
    Write the chemical reaction for the cell.

13. In the spaces below, write the values that will complete the expression for the potential of the following cell.

Cu |Cu²⁺(2.0M) || Ag⁺(1.5M) |Ag

(a) $-\dfrac{0.05916}{(b)} \log \dfrac{[(c)]^{(e)}}{[(d)]^{(f)}}$

a. _____

b. _____

c. _____

d. _____

e. _____

f. _____

14. Predict whether the potential of the cell in question 13 will be greater or less than that of the standard cell in question 12.

_____

15. In the spaces below, write the values that will complete the expression for calculating the mass of copper deposited by a current of 6.0 amperes flowing for a period of 1.00 min.

(a)	1 min	(b)	1C	1 mol e⁻	1 mol Cu	(d) Cu
		1 min	A·s	96 485 C	(c) mol e⁻	1 mol Cu

a. _____

b. _____

c. _____

d. _____

Chapter 27
STUDY GUIDE
27.1 INTRODUCTORY THERMODYNAMICS

1. For each of the following processes, determine whether the indicated quantities of the system are greater than zero, equal to zero, or less than zero. Write your answer in the blank following each quantity. If not enough information is given to determine a value, place an X in the blank.

   a. A system loses heat.
      q _____   w _____ ΔU _____

   b. A sample of a gas is compressed without heat transfer.
      q _____   w _____ ΔU _____

   c. A system does work without a decrease in its internal energy.
      q _____   w _____ ΔU _____

   d. A sample of a gas is heated. As the gas expands, it does work pushing a piston.
      q _____   w _____ ΔU _____

   e. The internal energy of a confined gas increases without work being done.
      q _____   w _____ ΔU _____

Figure 1 is a pressure-temperature graph of a system showing two paths (arrows) that represent independent ways of changing the system. The internal energy of the system is given by $\Delta U = q + w$. Use Figure 1 to answer questions 2-9.

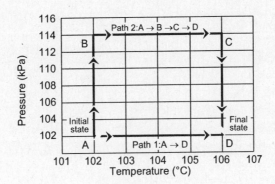

Figure 1

2. Underline each of the following variables that is a state function.

   P   T   q   w ΔU

3. Underline each of the following paths that represents an isobaric process.

   A→B   B→C   C→D   A→D

4. Underline each of the following paths that represents an isothermal process.

   A→B   B→C   C→D   A→D

5. What is the final temperature of the system?

   _____

6. What is the final pressure of the system?

7. What is ΔT of the system?

8. What is ΔP of the system?

9. If ΔH of the system is 20.0 J, determine q_p

Figure 2 represents the changes of enthalpy of the reactions
$$H_2(g) + \frac{1}{2}O_2(g) \rightarrow H_2O(l)$$
$$H_2O(l) \rightarrow H_2O(g)$$

Use Figure 2 to answer questions 10 - 14.

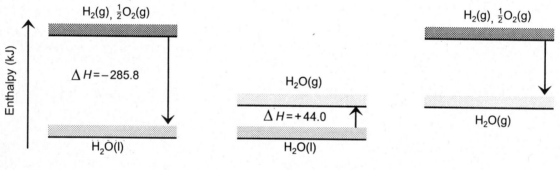

Figure 2

10. Which of the two reactions represents only a physical change?

11. Which reactants have enthalpies of formation of 0 kJ/mol?

12. Which reactions are exothermic?

13. Write the chemical equation for the formation of $H_2O(g)$ from its elements.

14. Determine the enthalpy of formation of $H_2O(g)$ from its elements.

Chapter 27
STUDY GUIDE
27.2 DRIVING CHEMICAL REACTIONS

For each of the following situations determine if the entropy change will be positive (+) or negative (−). Write a plus or minus sign in the blank provided.

___ 1. Calcium chloride crystals are pulverized in a mortar and pestle.

___ 2. Turpentine evaporates.

___ 3. A mixture of iron and sulfur is separated using a magnet.

___ 4. A spark plug ignites a fuel-air mixture in an engine.

___ 5. Turpentine dissolves a paint stain.

___ 6. Melted solder solidifies.

___ 7. Air is heated.

___ 8. Graphite is recrystallized as diamond in a press.

9. Complete the table by naming each state function and giving the appropriate SI unit for a molar quantity.

Symbol	State function	SI unit
ΔH_f°		
ΔG_f°		
ΔS°		
E°		

Answer the following questions.

10. What are standard states of temperature and pressure for measuring thermodynamics quantities?

11. What symbol is used to indicate that a quantity has been measured under standard states?

12. What is characteristic about the Gibbs free energies of reactions that are spontaneous?

13. What is characteristic about the Gibbs free energies of reactions that are at equilibrium?

14. What is characteristic about the Gibbs free energies of reactions that are not spontaneous?

Place a check mark (✓) in the blank next to each set of conditions that tends to favor spontaneous reactions. Place an X in the blank if the conditions do not favor spontaneous reactions.

		ΔH	T	ΔS			ΔH	T	ΔS
___	15.	<0	low	<0	___	19.	>0	low	<0
___	16.	<0	low	>0	___	20.	>0	low	>0
___	17.	<0	high	<0	___	21.	>0	high	<0
___	18.	<0	high	>0	___	22.	>0	high	>0

23. Use the information in Table 1 to complete Table 2 for the following reaction that takes place under standard measurement conditions.

$$CaCO_3(cr) \rightarrow CaO(cr) + CO_2(g)$$

Table 1

State function	Compound		
	$CaCO_3(cr)$	$CaO(cr)$	$CO_2(g)$
$\Delta G_f°$ (kJ/mol)	−1.129	−0.641	−1.650
$S°$ (J/mol·K)	92.8	39.7	885.6

Table 2

Quantity	Value	Quantity	Value
$\Sigma \Delta G_f°_{(products)}$		$\Sigma S_f°_{(products)}$	
$\Sigma \Delta G_f°_{(reactants)}$		$\Sigma S_f°_{(reactants)}$	
$\Delta G_r°$		$S_r°$	

24. Is the reaction in question 23 spontaneous? Explain your reasoning.

_____

Chapter 28
STUDY GUIDE

28.1 NUCLEAR STRUCTURE AND STABILITY

Answer each question.

1. Name two devices that produce high-energy particles used to bombard nuclei in the investigation of nuclear structure, and compare how the devices are used.

2. List three things that may happen when a nucleus is bombarded by a high-energy particle, and describe the possible effect of each on the bombarded nucleus.

3. On what three things does the type of change of a bombarded nucleus depend?

4. State two problems associated with the use of nuclear reactors to produce energy for practical use.

5. Radioactive element X decays to element Y. A sample initially contains 72 g of element X. At the end of three half-lives, how many grams of element X does the sample contain?

Write T for true or F for false. If a statement is false, replace the underlined word or phrase with one that will make the statement true, and write your correction on the blank provided.

_____ 6. In a synchrotron, <u>drift tubes</u> are used to confine fast moving particles to the ring.

_____ 7. In a <u>circular</u> accelerator, electromagnetic waves are used to accelerate particles to optimum speed.

_____ 8. Neutrons easily penetrate and can be absorbed by nuclei because the neutrons have <u>a negative charge</u>.

_____ 9. <u>Fission</u> involves the breaking apart of a heavy nucleus into two parts of about the same mass.

_____ 10. The emission and absorption of <u>alpha particles</u> are necessary in sustaining a chain reaction.

_____ 11. In nuclear reactions, the law of conservation of <u>mass-energy</u> is upheld.

_____ 12. In a nuclear reactor, a <u>moderator</u> is a device that regulates the rate at which the chain reaction takes place.

_____ 13. The half-life of a radioactive element is the amount of time it takes for half the atoms of a sample to <u>react with other atoms</u>.

_____ 14. A sample of a radioactive substance has a mass of 10 g. At the end of two half-lives, <u>0 g</u> of the original sample remains.

_____ 15. Because it does not absorb neutrons, water is often used as a <u>coolant</u> in nuclear reactors.

_____ 16. A <u>synchrotron</u> is a device for controlling fission.

_____ 17. The <u>containment vessel</u> of a nuclear reactor protects people from the heat and radiation of the reactor core.

Match each substance with the part of a nuclear reactor in which the substance is used. Some substances may be used in more than one part of the reactor.

a. moderator
b. control rods
c. coolant
d. fuel
e. containment vessel

___ 18. water

___ 19. uranium-235

___ 20. cadmium

___ 21. plutonium-239

___ 22. molten sodium

___ 23. helium

___ 24. graphite

___ 25. boron

26. Complete the table below to show how atomic number and mass number are affected in various transmutations.

Decay mode	Effect on atomic number	Effect on mass number
α-particle		
	increases by 1	
	decreases by 1	

27. Label the parts of the nuclear reactor shown in the figure and describe the function of each part.

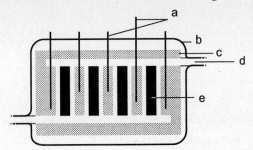

a. _____
b. _____
c. _____
d. _____
e. _____

Use Figure 28.6 in your textbook to answer the following questions about the nuclear decay of uranium-238 to lead-206.

28. What is the total number of alpha particles emitted during the process?
   _____

29. What is the total number of beta particles emitted?
   _____

30. What is the atomic number of the nuclide formed at the end of step 5?
   _____

31. What is the mass number of the nuclide formed at the end of step 8?
   _____

32. What is the name of the nuclide formed at the end of step 9?
   _____

33. What is the name of the nuclide with the shortest half-life?
   _____

34. How is the nuclide produced in step 11 different from the nuclide produced at the end of the decay series?
   _____
   _____

Chapter 28

STUDY GUIDE

28.2 NUCLEAR APPLICATIONS

Write T for true or F for false. If a statement is false, replace the underlined word or phrase with one that will make the statement true, and write your correction on the blank provided.

_____ 1. Elements that have atomic numbers greater than 92 are known as the <u>transmutation</u> elements.

_____ 2. Because of their relatively large mass and charge, the particles that, once inside the body, are most damaging to cells are <u>beta particles</u>.

_____ 3. Alpha particles are the <u>least</u> penetrating type of radiation.

_____ 4. Radioactive elements are useful as tracers because they can be easily <u>disintegrated</u>.

_____ 5. In <u>fission</u>, the nuclei of small atoms unite to form a larger nucleus.

_____ 6. Fusion generally produces a <u>greater</u> amount of energy per particle than does fission.

_____ 7. Carbon-14 dating is most useful for determining the age of <u>rocks</u>.

_____ 8. Radiation damage to living cells is indicated by the amount of <u>emitted</u> radiation.

_____ 9. Transmutations resulting from "packing" neutrons into the nuclei of certain elements produce elements with atomic numbers <u>lower</u> than those of the original elements.

Answer each question.

10. List three ways in which transuranium elements may be produced.

_____

_____

11. Differentiate between a gray and a sievert.

_____

_____

12. List four sources of radiation to which humans are generally exposed.

_____

_____

_____

13. List four ways in which radioactive nuclides are put to practical use.

_____

_____

14. State three problems associated with the development of plasma fusion reactors.

_____

_____

_____

Chapter 29
STUDY GUIDE
29.1 HYDROCARBONS

Complete each sentence.

1. A chain compound in which all carbon-carbon bonds are single is called an _____ or a _____.

2. Each alkane differs from the next by a(n) _____ group.

3. With increasing molecular mass of compounds within a homologous series, the boiling point _____.

Write the name that corresponds with the following chemical formulas of compounds or radicals.

4. C_5H_{11} _____

5. C_8H_{18} _____

6. C_3H_8 _____

7. C_4H_9 _____

Write T for true or F for false. If a statement is false, replace the underlined word or phrase with one that will make the statement true, and write your correction on the blank provided.

_____ 8. In <u>a branched</u> chain molecule, the numbering of carbon atoms can begin at either end of the chain.

_____ 9. When a compound has more than one branch, radicals appear in <u>numerical order</u> in the name of the compound.

_____ 10. <u>Prefixes</u> are used in the naming of compounds in which two or more substituent groups are alike.

_____ 11. The term *trimethyl* means <u>one three-carbon branch</u>.

_____ 12. Isomers <u>are not</u> named according to the total number of carbon atoms in a molecule.

Match the general formulas with the corresponding class of organic compounds.

a. ester
b. ether
c. ketone
d. acid
e. alcohol

___ 13. R–C(=O)–O–H

___ 14. R–C(=O)–O–R'

___ 15. R–O–H

___ 16. R–C(=O)–R'

___ 17. R–O–R'

STUDY GUIDE 119

Name _____ Date _____ Class _____

Define the following terms.

18. cycloalkanes _____
_____

19. alkenes _____
_____

20. unsaturated hydrocarbons _____
_____

21. alkynes _____
_____

22. aromatic hydrocarbons _____
_____

23. How does a benzene ring differ from cyclohexane?
_____
_____
_____

Draw the structural formulas for the following compounds in the spaces provided.

24. benzene

25. cyclohexane

26. phenyl radical

27. methylcyclohexane

28. 1,3-dimethylcyclopentane

120   STUDY GUIDE

29. toluene

30. 2,2,4-trimethylpentane

31. 1-methyl-3-ethylcyclohexane

Answer each of the following.

32. What do benzene, toluene, and xylene have in common?
    _____

33. What materials is styrene used to make?
    _____

Name _____ Date _____ Class _____

Chapter 29
STUDY GUIDE

29.2 OTHER ORGANIC COMPOUNDS

Match the corresponding terms by choosing from the list below.

a. chloromethane
b. methanol
c. 1,2-dichloroethane
d. chloroethene
e. trichloromethane
f. 1,2-ethanediol

___ 1. is also known as chloroform

___ 2. has the formula CH_3Cl

___ 3. is also known as vinyl chloride

___ 4. is also known as ethylene dichloride

___ 5. has the formula CH_3OH

___ 6. is also known as ethylene glycol

Write T for true or F for false. If a statement is false, replace the underlined word or phrase with one that will make the statement true, and write your correction on the blank provided.

_____ 7. Because aromatic hydroxyl compounds have properties that differ somewhat from those of most alcohols, they are classified as <u>bases</u>.

_____ 8. Ethoxyethane was used for many years as <u>an antiseptic</u>.

_____ 9. Alcohols are <u>neither acidic or basic</u>.

_____ 10. Alcohols with four or more carbon atoms <u>are soluble</u> in water.

Fill in the missing term(s).

11. The most important ketone in industry is _____.

12. Methanal is more commonly known as _____.

13. Both aldehydes and ketones are characterized by the presence of a(n) _____ group.

14. The effect of one functional group on another is called a(n) _____.

15. Because they do not ionize greatly in water, most organic acids are _____ acids.

16. The _____ acid group characterizes most organic acids.

17. Acetic anhydride is made from _____ by the removal of a molecule of water.

18. In the reaction between ethanol and acetic acid, _____ and _____ are produced.

122 STUDY GUIDE Chapter 2

Name _____ Date _____ Class _____

Define the following terms.

19. amine _____

20. primary amine _____

21. amide _____

22. nitrile _____

Write the general formula for each of the following compounds.

23. nitro compounds

24. amines

25. nitriles

Chapter 29 STUDY GUIDE **123**

Name _____ Date _____ Class _____

Chapter 30
STUDY GUIDE

30.1 Organic Reactions

Match the reactants in questions 1-6 with the following products. Then identify the type of organic reaction illustrated in each question.

a. H-C(H)(H)-C(H)(H)-H + HCl (with benzene ring on first C)

b. H-C(H)(H)-C(H)(H)-OH

c. H-C(H)(=O)-O-C(H)(H)-C(H)(H)-H + H_2O

d. H-C(H)(H)-C(H)(I)-C(H)(H)-H

e. H-C(H)(H)-C(H)=C(H)(H) + H_2O

f. H-C(H)(H)-H (with benzene ring) + CH_3Cl

g. -C(H)(Cl)-C(H)(H)-C(H)(Cl)-C(H)(H)-C(H)(Cl)-C(H)(H)-

_____ 1. H-C(H)(H)-C(H)=C(H)(H) + HI →

_____ 2. (benzene) + H-C(H)(H)-C(H)(H)-Cl $\xrightarrow{AlCl_3}$

_____ 3. H-C(H)(H)-C(H)(OH)-C(H)(H)-H $\xrightarrow{H_2SO_4}$

_____ 4. H(H)C=C(H)H + H_2O $\xrightarrow{\text{dilute } H_2SO_4}$

_____ 5. H-C(H)(=O)-OH + H-C(H)(H)-C(H)(H)-OH →

_____ 6. H(H)C=C(H)(Cl) + H(H)C=C(H)(Cl) + H(H)C=C(H)(Cl) + ... →

Name _____ Date _____ Class _____

Write T *for true or* F *for false. If a statement is false, replace the underlined word or phrase with one that will make the statement true, and write your correction on the blank provided.*

_____ 7. In a condensation polymerization reaction, one functional group may be replaced by another.

_____ 8. In an addition reaction, a double bond between carbons is created.

_____ 9. A reaction in which water is removed from an alcohol is an esterification reaction.

_____ 10. Soap is made by the polymerization of fatty acids.

_____ 11. Like molecules called *monomers* chain together to form polymers.

_____ 12. When an ester is deformed by an outside force, it returns to its original shape when the force is removed.

_____ 13. Rayon is made from reconstituted cellulose.

_____ 14. Many compounds containing benzene readily take part in addition reactions.

_____ 15. In the fractional distillation of petroleum, the boiling point of the gasoline fractions must be lower than the boiling point of the fuel oil fractions.

_____ 16. Cracking is a process used to increase the yield of natural rubber.

_____ 17. A thermoplastic material hardens upon heating.

_____ 18. The higher the octane rating, the greater the percent of the fuel that burns evenly.

Chapter 30
STUDY GUIDE
30.2 BIOCHEMISTRY

Write T for true or F for false. If a statement is false, replace the underlined word or phrase with one that will make the statement true, and write your correction on the blank provided.

_____ 1. Enzymes are <u>lipid</u>-based biological catalysts.

_____ 2. Proteins are composed of <u>nucleic acids</u> linked by peptide bonds.

_____ 3. Steroids, some vitamins, and fats are classified as <u>carbohydrates</u>.

_____ 4. <u>Carbohydrates</u> are more soluble in water than in nonpolar solvents.

_____ 5. Protein synthesis in the human body is controlled by <u>amino acids</u>.

_____ 6. Polysaccharides are formed by <u>addition</u> polymerization.

_____ 7. Cartilage and tendons are composed mainly of <u>cellulose</u>.

Answer the following questions.

8. State three ways in which proteins may differ from one another.

_____

_____

_____

9. List three glucose-based polysaccharides found in living things.

_____

10. State three problems associated with the use of biomaterials.

_____

_____

_____

11. List four uses of biomaterials.

_____

_____

12. Compare the chemical composition of RNA with that of DNA.

_____

_____

_____

Refer to Table 30.1 in your textbook to answer the following questions.

13. Draw a structural formula for the amino acid serine.

14. Draw a structural formula for the amino acid glycine.

15. Use structural formulas to show how serine and glycine may combine to form two different dipeptides.

Chapter 1
STUDY GUIDE

THE ENTERPRISE OF CHEMISTRY

Complete each sentence.

1. Matter is anything that has the property of __inertia__.

2. The way that matter behaves is described by its __properties__.

3. The property possessed by all matter, which in the proper circumstances can be made to do work, is called __energy__.

4. A property of matter that shows itself as a resistance to any change in motion is called __inertia__.

5. Potential energy depends upon the __position__ of an object with respect to some reference point.

6. Kinetic energy is the energy possessed by an object because of its __motion__.

7. Energy that is transferred as electromagnetic waves is called __radiant__ energy.

Answer each question.

8. What do chemists do?
   Chemists solve problems that have to do with matter; they discover the relationships between the structure and properties of matter and use the knowledge to produce new materials.

9. What are the two ways energy can be transferred between objects?
   through direct contact and through electromagnetic waves

10. What does the law of conservation of mass state?
    Matter is neither created nor destroyed, but is changed only in form.

11. What does the law of conservation of energy state?
    Energy is neither created nor destroyed, but is changed only in form.

12. What does the law of conservation of mass-energy state?
    Mass and energy can be changed from one to the other, but their sum remains constant.

13. When are changes of energy to mass and mass to energy observable?
    in nuclear reactions

14. What is an intermediate?
    An intermediate is a material used in the production of consumer goods that is neither a raw material nor a consumer product.

15. Why would water used in the making of a consumer product be considered a raw material and not an intermediate?
    Water is a naturally occurring substance on Earth.

16. What is a model?
    A model is a picture or representation that is used to help deal with abstract ideas and objects.

17. What law does Einstein's equation $E = mc^2$ express?
    the law of conservation of mass-energy

18. Explain why pesticides did not succeed in permanently controlling insect populations.
    Some individuals were resistant to it and multiplied rapidly after the susceptible population was killed off.

19. State two harmful effects of pesticides.
    They may remain in the environment and be taken in and stored by animals for which they were not intended. They may kill good insects as well as harmful ones.

Chapter 2
STUDY GUIDE

2.1 DECISION MAKING

Answer each of the following.

1. What are the two requirements that must be met before the answer to a chemistry problem can be called correct?

   The answer must be numerically correct and have the correct units.

2. In solving problems, what should you do if the problem includes irrelevant information?

   Ignore the irrelevant information.

3. Why should you carefully consider the units in which your answer will be expressed?

   To see if your solution will give those units

4. Why is it important to examine your calculated answer to a problem?

   To see if the answer is reasonable and makes sense

5. In the table below, list the seven problem-solving steps outlined in your text. Then, read the problem and describe how to apply each of the seven steps to solve the problem.

   Problem: Seven students are working to put together a student newsletter. There are six sheets of paper in each newsletter. Three students can each assemble two newsletters per minute. Three other students can each assemble three newsletters per minute. The remaining student can assemble four newsletters per minute. How many newsletters will the group assemble in one hour?

Steps	Application of steps
(1) Identify the known facts.	number of students, rate of each; note that number of sheets is irrelevant
(2) Define the answer required.	newsletters per hour
(3) Develop possible solutions.	Find hourly rate for each student, then add; or find rate for the group, then find hourly rate; incorrect solutions also can be considered at this step.
(4) Analyze these solutions and choose the correct one.	Answers will vary, but a correct solution should be chosen.
(5) Develop the individual steps to arrive at the answer.	Answers will vary depending on method chosen.
(6) Solve the problem.	The correct answer is 1140 per hour.
(7) Evaluate the results.	Results should seem reasonable.

Chapter 2
STUDY GUIDE

2.2 NUMERICAL PROBLEM SOLVING

1. Distinguish between quantitative and qualitative descriptions.

   Quantitative descriptions include measurements; qualitative descriptions do not.

2. Complete the table.

Quantity	Unit	Unit symbol
Electric current	ampere	A
Length	meter	m
Time	second	s
Mass	kilogram	kg
Temperature	kelvin	K
Amount of substance	mole	mol
Luminous intensity	candela	cd

3. List the following units from largest to smallest: meter, millimeter, kilometer, centimeter, picometer.

   kilometer, meter, centimeter, millimeter, picometer

Match each quantity to be measured with the most appropriate SI unit.

a. centimeter
b. gram
c. kelvin
d. kilogram
e. meter
f. newton
g. second

- e 4. length of a swimming pool
- a 5. length of a pencil
- b 6. mass of a pencil
- d 7. mass of a bag of apples
- g 8. time between two heartbeats
- c 9. temperature of boiling water
- f 10. weight of a textbook

11. Each of five students used the same ruler to measure the length of the same pencil. These data resulted: 15.33 cm, 15.34 cm, 15.33 cm, 15.34 cm, 15.55 cm. Using these numbers as examples, distinguish between accuracy and precision.

    Because the five measurements are very close, they can be described as precise. However, they are not very close to the actual value, so they are not accurate.

Chapter 3
STUDY GUIDE

3.1 CLASSIFICATION OF MATTER

Answer each of the following.

1. How does a material differ from a mixture?
   A material is a specific kind of matter. A mixture is matter that contains two or more different materials.

2. Describe an iceberg afloat in an ocean, using the terms *phase, state,* and *system*.
   The iceberg, which is in a solid state, and salt water, which is in the liquid state, are separate phases of the iceberg-saltwater system.

3. Compare and contrast heterogeneous and homogeneous mixtures and give two examples of each.
   A heterogeneous mixture is composed of more than one phase. Granite and milk are heterogeneous mixtures. Homogeneous mixtures are mixtures consisting of one phase. Salt water and air are homogeneous mixtures.

4. Suppose a glass contains 2 cups of sugar dissolved in 4 cups of water. Classify this mixture. Which component is the solute? Which is the solvent?
   It is a homogeneous mixture, or solution. The sugar is the solute and the water is the solvent.

5. Explain why sea water is classified as a solution.
   The composition of sea water varies from place to place, yet each sample is homogeneous.

6. What is a substance? How does it differ from a homogeneous mixture?
   A substance is a homogeneous material that always has the same composition. A homogeneous mixture, on the other hand, varies in composition from sample to sample.

7. How are elements and compounds related?
   Elements are matter made of only one kind of atom. Compounds are substances made of more than one kind of atom.

8. Are compounds more or less ordered than heterogeneous mixtures?
   Compounds are more ordered than heterogeneous mixtures.

9. Classify the following materials as heterogeneous mixtures, compounds, elements, or solutions: sugar, salt, water, gold, brass, wood, carbon, and air.
   sugar, salt, water: compounds; gold: element; brass: solution; wood: heterogeneous mixture; carbon: element; air: solution

10. Classify the following substances as organic or inorganic: methane (CH_4), aluminum chloride ($AlCl_3$), ethanol (C_2H_5OH), benzene (C_6H_6), and copper (Cu).
    organic: methane, ethanol, benzene; inorganic: aluminum chloride, copper

---

12. For each group of digits indicated in the three numbers below, state whether the digits are significant. Then state the rule that applies.

    $$\underset{a\ b\ c}{4500.60} \quad \underset{d\ e}{0.000\ 799} \quad \underset{f}{2\hat{2}0}$$

    a. Significant; all nonzero digits are significant.
    b. Significant; all zeros between two other significant digits are significant.
    c. Significant; one or more final zeros used after a decimal point are significant.
    d. Not significant; zeros used solely for spacing the decimal point (placeholders) are not significant.
    e. Significant; all nonzero digits are significant.
    f. Not significant; placeholder zeros are not significant.

13. A stack of books contains 10 books, each of which is determined, by a ruler graduated in centimeters, to be 25.0 cm long. How do these two quantities differ in terms of significant digits?
    Here the number 10 has an infinite number of significant digits because it expresses a natural-number count (number of books). The number 25.0 has three significant digits because it is a measurement and the value has been measured and reported to that degree of certainty.

14. What is the percent error in a determination that yields a value of 8.38 g/cm³ as the density of copper? The literature value for this quantity is 8.92 g/cm³.

    $$\frac{|8.38 \text{ g/cm}^3 - 8.92 \text{ g/cm}^3|}{8.92 \text{ g/cm}^3} \times 100 = 6.1\%$$

15. A student measures the melting point of ammonium acetate, $NH_4C_2H_3O_2$, as 117°C, but the literature value is 114°C. What is the percent error in the measurement?

    $$\frac{|117°C - 114°C|}{114°C} \times 100 = 3\% \text{ (rounded)}$$

Derive units that would be appropriate for each of the following situations.

16. A faucet trickles, even when it is shut off. How would you express the rate of water flow?
    mL/s or cm³/s

17. When you add ice cubes to a glass of water, the water's temperature decreases. How would you express the change in temperature caused by the addition of each ice cube?
    K/cube or C°/cube

18. Complete the following table of densities, using Table 2.3 in your text to determine the probable identity of X, Y, and Z.

Material	Mass (g)	Volume (cm³)	Density (g/cm³)	Identity
X	16.0	2.0	8.0	stainless steel
Y	17.8	2.0	8.9	copper
Z	2.7	3.4	0.79	ethanol

Name _____ Date _____ Class _____

Chapter 3 STUDY GUIDE

3.2 CHANGES IN PROPERTIES

Write T for true and F for false. If a statement is false, change the underlined word or phrase and write your correction on the blank.

__F; physical__ 1. A change in which the same substance remains after the change is called a <u>chemical</u> change.

__T__ 2. Melting is a <u>physical</u> change.

__F; physical__ 3. Evaporating water from a saltwater solution is a <u>chemical</u> change.

__F; solid__ 4. A <u>gaseous</u> substance that forms from a solution is called a precipitate.

__T__ 5. The reaction of iron with hydrochloric acid is a <u>chemical</u> change.

Use the figure to answer the following questions.

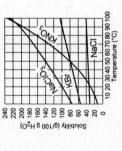

6. What is the solubility of sodium chloride at 30°C? __about 32 g/100 g H₂O__

7. Which is more soluble at 90°C: KNO₃ or NaClO₃? __NaClO₃__

8. At what temperature are the solubilities of KNO₃ and KBr the same? __about 50°C__

9. At what temperature will 200 g of NaClO₃ dissolve in 100 g of H₂O? __about 86°C__

Identify each property below as either a chemical property or a physical property. For each physical property, state whether it is extensive or intensive.

10. color __physical, intensive__   15. conductivity __physical, intensive__

11. width __physical, extensive__   16. luster __physical, extensive__

12. melting point __physical, intensive__   17. flammability __chemical__

13. density __physical, intensive__   18. weight __physical, extensive__

14. resistance to an acid __chemical__

Name _____ Date _____ Class _____

Chapter 3 STUDY GUIDE

3.3 ENERGY

Answer the following questions.

1. Recall that ordinary salt is a compound of sodium and chlorine. List several systems that can be defined in a shaker full of salt.

   Answers will vary, but might include: the shaker; the shaker and the salt; the salt alone; a number of crystals of salt; and one sodium atom and one chlorine atom.

2. What is heat?
   the energy transferred due to a temperature difference

3. What is a joule?
   a quantitative unit of energy or energy change

4. How is the calorie related to the joule?
   One calorie is equivalent to 4.184 joules.

5. Contrast endothermic and exothermic reactions.
   In endothermic reactions, energy is absorbed. In exothermic reactions, energy is released.

6. What is specific heat?
   the heat needed to raise the temperature of one gram of a substance one Celsius degree

Solve the following problems.

7. How many joules are in 1.11 Calories?
   (1.11 cal)(4.184 J/cal) = 4.64 J

8. An average baked potato contains 164 Calories. Calculate the energy value of the potato in joules.
   (164 Cal)(1000 cal/Cal)(4.184 J/cal) = 686 000 J

9. How much heat is lost when a 4110-g metal bar whose specific heat is 0.2311 J/g·C° cools from 100.0°C to 20.0°C?
   $q = (m)(\Delta T)(C_p)$
   = (4110 g)(80.0°C)(0.2311 J/g·C°)
   = 76 000 J

Chapter 4
STUDY GUIDE

4.1 EARLY ATOMIC MODELS

Complete the following sentences.

1. The cathode-ray tube _____ is an apparatus that helped scientists discover that atoms contain electrons.

2. Isotopes of an element have the same number of protons but different numbers of neutrons _____.

3. The number of protons in the nucleus of an atom is called the atomic number _____ of that element.

4. The total number of protons and neutrons in a nucleus is called the mass number.

5. The spontaneous production of rays of particles and energy by unstable nuclei is called radioactivity.

Match the names of the scientists to their discoveries.

a. Einstein
b. Proust
c. Becquerel
d. Marie and Pierre Curie
e. Geiger, Marsden, and Rutherford
f. Millikan
g. Thomson
h. Lavoisier
i. Dalton
j. Chadwick

__d__ 6. ability of radium to give off rays
__e__ 7. positively charged nucleus
__a__ 8. $E = mc^2$
__g__ 9. electron's charge-to-mass ratio
__f__ 10. electron's charge
__j__ 11. neutrons
__h__ 12. law of conservation of mass
__c__ 13. ability of uranium to expose photographic film
__i__ 14. law of multiple proportions
__b__ 15. law of definite proportions

---

10. What is the specific heat of copper if a 105-g sample absorbs 15 200 J and the change in temperature is 377°C?

    $C_p = q/(m)(\Delta T)$
    = 15 200 J/(105 g)(377°C)
    = 0.384 J/g·C°

11. What is the final temperature of a system of iron and water if a piece of iron with a mass of 25.0 grams at a temperature of 75.0°C is dropped into an insulated container of water at 35.5°C? The mass of the water is 150.0 grams. The specific heat of iron is 0.449 J/g·C°, and the specific heat of water is 4.184 J/g·C°.

    The heat lost by the iron is
    $q = (m)(\Delta T)(C_p)$
    = (25.0 g)(75.0°C − T_f)(0.449 J/g·C°)

    The heat gained by the water is
    $q = (m)(\Delta T)(C_p)$
    = (150.0 g)(T_f − 35.5°C)(4.184 J/g·C°)

    The heat gained must equal the heat lost, so
    (25.0 g)(75.0°C − T_f)(0.449 J/g·C°) = (150.0 g)(T_f − 35.5°C)(4.184 J/g·C°)

    Solving for T_f:
    11.2 J/C°(75.0°C − T_f) = 628 J/C°(T_f − 35.5°C)
    840 J − (11.2 J/C°) T_f = (628 J/C°) T_f − 22 300 J
    −(11.2 J/C°) T_f − (628 J/C°) T_f = −22 300 J − 840 J
    (−639.2 J/C°) T_f = −23140 J

    $T_f = \dfrac{-23\ 140\ J}{-639.2\ J/C°}$

    $T_f = 36.2°C$

Chapter 4
STUDY GUIDE

4.2 PARTS OF THE ATOM

Write T for true or F for false. If a statement is false, replace the underlined word or phrase with one that will make the sentence true, and write your correction on the blank provided.

T _____ 1. Alpha radiation and beta radiation are made up of <u>particles</u>.

F: high _____ 2. Gamma radiation is of very <u>low</u> energy.

T _____ 3. Waves that are slightly higher in frequency than visible light are called <u>ultraviolet</u> rays.

F: lambda (λ) _____ 4. Wavelength is represented by the Greek letter <u>nu (ν)</u>.

F: light _____ 5. Electromagnetic energy travels at the speed of <u>sound</u>.

T _____ 6. Electromagnetic waves with low frequencies have <u>long</u> wavelengths.

F: Emission _____ 7. <u>Absorption</u> spectra are produced by energy given off by excited gaseous atoms.

T _____ 8. The Rutherford-Bohr model of the atom is sometimes called the <u>planetary</u> model.

F: $E = h\nu$ _____ 9. Planck proposed the equation $E = mc^2$.

T _____ 10. A <u>photon</u> is a quantum of radiant energy.

Match the particles to the descriptions or examples.

a. beta particle   f. lepton
b. photon   g. quarks
c. alpha particle   h. nucleon
d. baryon   i. antiparticle
e. meson   j. gluon

j _____ 11. particle exchanged by quarks
c _____ 12. helium nucleus
f _____ 13. an electron, neutrino, muon, or pion
e _____ 14. always contains a quark and an antiquark
i _____ 15. a positron, for example
a _____ 16. an electron released by a nucleus
d _____ 17. always contains three quarks
b _____ 18. energy packet of light
h _____ 19. any component of a nucleus
g _____ 20. "up," "down," "charm," "strange," "top," or "bottom"

---

16. Briefly state the following:

a. law of definite proportions

Specific compounds always contain elements in the same ratio by mass.

b. law of multiple proportions

The ratio of masses of one element that combines with a constant mass of another element to form more than one compound can be expressed in small whole numbers.

17. Complete the following table of isotopes of oxygen.

Isotope	Protons	Neutrons	Mass number
oxygen-16	8	8	16
oxygen-17	8	9	17
oxygen-18	8	10	18

18. What is wrong with the diagram of a cathode-ray tube? How should it look? Explain your answer.

The ray should not continue straight to the right, but should be bent by the magnet. This bending occurs because the ray is composed of negatively charged electrons, whose path is affected by a magnetic field.

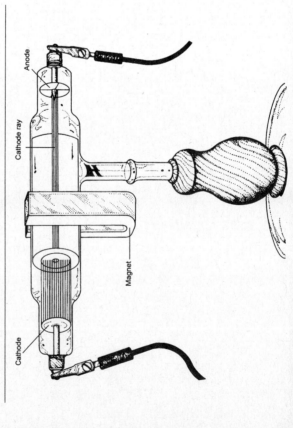

Name _____ Date _____ Class _____

Chapter 5
STUDY GUIDE

5.1 MODERN ATOMIC STRUCTURE

Complete the sentence or answer the question.

1. Define or explain the following.

   a. wave-particle duality of nature   Waves can act as particles, and particles can act as waves.

   b. Newtonian mechanics   classical mechanics that describes the behavior of visible objects traveling at ordinary velocities

   c. quantum mechanics   mechanics that describes the behavior of extremely small particles traveling at velocities near that of light

   d. the Heisenberg uncertainty principle   states that it is impossible to know both the exact position and the exact momentum of an object, such as an electron, at the same time

   e. quantum numbers   numbers used in describing behavior of an electron in an atom

2. The product of mass and velocity of an object is called the   momentum   .

3. The French scientist whose hypothesis on the wave nature of particles helped lead to the present-day theory of atomic structure was   de Broglie   .

4. The region in which an electron travels around the nucleus of an atom is called a(n)   electron cloud   .

5. Complete the table, using de Broglie's equation, $\lambda = \dfrac{h}{mv}$, for the wavelength λ of a particle of mass m and velocity v. Note: Planck's constant $h = 6.63 \times 10^{-34}$ kg·m²/s

m (kg)	v (m/s)	λ (m)
0.005 00	9.10×10^6	1.46×10^{-38}
0.0220	2.00×10^2	1.51×10^{-34}
1.15×10^{-29}	5.10	1.13×10^{-5}

6. Explain what is meant by the *probability* of finding an electron at a point in space.   the ratio between the number of times the electron is in that position divided by the total number of times it is at all possible positions

---

Name _____ Date _____ Class _____

Complete the following sentences.

21. Electromagnetic energy   includes visible light, radio waves, and infrared, ultraviolet, and X rays.

22. Alpha, beta, and gamma rays given off by nuclei are all forms of   radiation   .

23. Scientists use a nuclide of the element   carbon   as the standard for the atomic mass scale.

24. The symbol for an atomic mass unit is   u   .

25. The   average   atomic mass takes into account the differing masses and percent occurrences of isotopes of an element.

26. Inside a mass spectrometer, the paths of heavy particles are bent   less   than are the paths of lighter particles.

27. Calculate the average atomic mass of lithium, which occurs as two isotopes that have the following atomic masses and abundances in nature: 6.017 u, 7.30%; and 7.018 u, 92.70%.

    $\dfrac{(7.30 \times 6.017 \text{ u}) + (92.70 \times 7.018 \text{ u})}{100} = 6.94$ u

Chapter 5 STUDY GUIDE

5.2 QUANTUM THEORY

Complete the sentence or answer the question.

1. When an electron in a hydrogen atom moves from a higher to a lower energy state, the energy difference is emitted as a quantum of <u>light</u> .

2. Define the four quantum numbers n, l, m, and s, explain what information is given by each, and describe the range of values each may take.

   The *principal quantum number*, n, describes the energy level of an electron, and may take on any positive whole-number value.

   The *energy sublevel quantum number*, l, designates the geometrical shape of the region in space an electron occupies and the sublevel (s, p, d, or f) and may take on whole-number values from 0 to $n-1$.

   The *orbital quantum number*, m, designates the spatial orientation of the atomic orbital and may take on values from -1 to $+1$.

   The *spin quantum number*, s, describes the spin orientation of the electron and may take on values of $+1/2$ or $-1/2$.

3. Orbitals of the same energy are said to be <u>degenerate</u> .

4. The space occupied by one pair of electrons is called a(n) <u>orbital</u> .

5. What is the formula for calculating the maximum number of electrons that can occupy any energy level in an atom? <u>$2n^2$</u>

6. Complete the following table.

Energy level	Number of sublevels	Number of orbitals	Maximum number of electrons
1	1	1	2
2	2	4	8
3	3	9	18
4	4	16	32

7. State the *Pauli exclusion principle*.

   No two electrons in an atom have the same set of four quantum numbers.

Write T for true or F for false. If a statement is false, replace the underlined word or phrase with one that will make the statement true, and write your correction on the blank provided.

<u>F; Planck's quantum theory</u>   7. De Broglie used Einstein's relationship between matter and energy and <u>Schrödinger's wave equation</u> to develop his equation for wavelength of a particle.

<u>F; momentum</u>   8. To be able to give a full description of an electron, you would need to know two things: its present position and its <u>radius</u>.

<u>T</u>   9. It is <u>impossible</u> to know both the exact position and the exact momentum of an object at the same time.

<u>T</u>   10. One of the most important developments of <u>Schrödinger's wave equation</u> was the idea of quantum numbers, which describe the behavior of electrons.

<u>F; positive</u>   11. In Schrödinger's wave equation, n can have only <u>negative</u> whole number values.

Match each mathematical expression with the correct description.

a. de Broglie's equation for the wavelength of a particle
b. Einstein's equation relating matter and energy
c. Planck's quantum equation
d. term for total energy in Schrödinger's wave equation

<u>a</u>   12. $\lambda = \dfrac{h}{mv}$

<u>d</u>   13. $\dfrac{2\pi^2 me^4}{h^2 n^2}$

<u>c</u>   14. $E = h\nu$

<u>b</u>   15. $E = mc^2$

Chapter 5
STUDY GUIDE

5.3 DISTRIBUTING ELECTRONS

Complete the sentence or answer the question.

1. In using the "rule of thumb" arrow diagram to predict electron arrangements for most atoms in the ground state, what assumption must be made? _that the energy levels of electrons are not affected by the presence of other electrons_

2. It is important to know how to find the electron arrangement of an atom in order to predict its ___chemical reactivity___.

3. The number of electrons used in writing the electron configuration of an atom of an element is equal to the element's ___atomic number___.

4. Write the electron configurations for the following elements:
   a. lithium $1s^2 2s^1$
   b. boron $1s^2 2s^2 2p^1$
   c. sodium $1s^2 2s^2 2p^6 3s^1$
   d. sulfur $1s^2 2s^2 2p^6 3s^2 3p^4$
   e. calcium $1s^2 2s^2 2p^6 3s^2 3p^6 4s^2$

5. The notation in which outer-energy-level electrons are indicated around the symbol of an element is referred to as a(n) ___Lewis electron dot___ diagram.

6. Write the three steps for drawing electron dot diagrams.
   1. Write the symbol of the element and have it represent the nucleus and all electrons except those in the outer level.
   2. Write the electron configuration of the element, and select the electrons that are in the outermost energy level (those with the highest principal quantum number).
   3. Draw one dot around the symbol for each electron found in the outermost energy level.

7. Write the electron dot diagrams for the following elements:
   a. lithium   d. sulfur
      Li·   :S̈:
   b. boron   e. calcium
      :B·   Ca:
   c. sodium
      Na·

---

8. Complete the following table.

Sublevel	Number of orbitals	Maximum number of electrons
s	1	2
p	3	6
d	5	10
f	7	14

9. Complete the electron configurations for the following atoms by drawing in the arrows indicating the electrons with appropriate spins for each orbital.

            1s   2s   2p

   a. beryllium (atomic number 4) (↑↓) (↑↓)
   b. carbon (atomic number 6) (↑↓) (↑↓) (↑)(↑)( )
   c. fluorine (atomic number 9) (↑↓) (↑↓) (↑↓)(↑↓)(↑)

Write T for true or F for false. If a statement is false, replace the underlined word with one that will make the statement true, and write your correction on the blank provided.

__T__ 10. If two electrons occupy the same orbital, they must have __opposite__ spins.

__T__ 11. The __principal__ quantum number describes the energy level of an electron in an atom.

__F: one-electron__ 12. The Schrödinger wave equation is solvable for any __multielectron__ system.

__T__ 13. __Four__ quantum numbers are required to describe completely an electron in an atom.

__F: spherical__ 14. The sum of all electron clouds in any sublevel (or energy level) is a __tetrahedral__ cloud.

Chapter 6
STUDY GUIDE

6.1 DEVELOPING THE PERIODIC TABLE

Fill in the blanks with appropriate terms.

1. The table below shows the way a scientist named __Johann Dobereiner__ classified the elements into groups he called __triads__.

Symbol	Atomic mass	Symbol	Atomic mass
Ca	40.1	Cl	35.5
Ba	137.3	I	126.9
Ca-Ba Average	88.7	Cl-I Average	81.2
Sr	87.6		Br	79.9

2. The table below shows the classification of elements according to a principle called the __law of octaves__, which was developed by __John Newlands__.

1	2	3	4	5	6	7
Li	Be	B	C	N	O	F
Na	Mg	Al	Si	P	S	Cl

3. The elements in Mendeleev's table were arranged in order of increasing __atomic mass__.

4. As a result of Henry Moseley's work, the modern periodic table is arranged according to increasing __atomic number__.

5. The atomic number of an element indicates the number of __protons__ in the nucleus.

Write T for true and F for false. If a statement is false, replace the underlined word with one that makes the statement true.

__T__ 6. The electron configurations of hydrogen and helium are <u>not similar</u>, so each element is in a separate column of the periodic table.

__F; transition elements__ 7. The elements in columns 3 through 12 (IIIB through IIB) are called the <u>noble gases</u>.

__T__ 8. Each time a new principal energy level is started, a new <u>row</u> in the periodic table begins.

__T__ 9. The lanthanoid series contains the elements lanthanum through <u>ytterbium</u>.

__T__ 10. In the periodic table, a horizontal row of elements is called a <u>period</u>.

Answer the following questions.

11. What is the pattern of placing electrons in energy sublevels for elements in the actinoid series?
    Elements in the actinoid series generally have increasing numbers of electrons in the 5f sublevel.

12. Why are neon and helium placed in the same column in the periodic table?
    Helium has a completely filled first energy level, and neon has a completely filled second energy level.

13. Compare the ways Dobereiner and Newlands classified the elements.
    Dobereiner grouped elements into triads with similar properties. Newlands arranged elements in order of increasing atomic masses, in groups of seven.

14. Fill in the following table.

Sublevel type	Electron capacity
s	2
p	6
d	10
f	14

Chapter 6
STUDY GUIDE

6.2 USING THE PERIODIC TABLE

Match each of the following elements with the element in the list that has the most similar chemical properties. Use the periodic table as a guide.

a. zinc (Zn)
b. helium (He)
c. potassium (K)
d. rhodium (Rh)
e. terbium (Tb)
f. vanadium (V)
g. phosphorus (P)
h. gold (Au)
i. indium (In)
j. chlorine (Cl)
k. rhenium (Re)
l. oxygen (O)
m. yttrium (Y)
n. silicon (Si)
o. radium (Ra)

_d_ 1. cobalt (Co)
_n_ 2. carbon (C)
_b_ 3. argon (Ar)
_e_ 4. cerium (Ce)
_c_ 5. lithium (Li)
_i_ 6. aluminum (Al)
_j_ 7. iodine (I)
_h_ 8. copper (Cu)
_k_ 9. manganese (Mn)
_f_ 10. niobium (Nb)
_o_ 11. calcium (Ca)
_m_ 12. scandium (Sc)
_a_ 13. mercury (Hg)
_l_ 14. selenium (Se)
_g_ 15. arsenic (As)

Write a T for true or F for false. If the statement is false, replace the underlined word with a word that makes the statement true.

T 16. The electron configurations of all elements in Group 1 (IA) end in s^1.

F; period number 17. In the electron configuration for the outer energy level of potassium, $4s^1$, the coefficient 4 indicates the <u>group number</u>.

F; nonreactive 18. Atoms with full outer energy levels are <u>very reactive</u>.

T 19. Elements such as silicon are called metalloids because they have properties of <u>both metals and nonmetals</u>.

T 20. Elements with three or fewer electrons in the outer energy level are usually <u>metals</u>.

Answer each of the following.

21. Describe the properties of nonmetals, and give examples.

Nonmetals are insulators, with dull surfaces. They are usually gases or brittle solids at room temperature. Examples may include most elements from groups 16 (VIA), 17 (VIIA), and 18 (VIIIA).

22. What is the main reason that atoms react with each other?

Atoms react because they become more stable after the reaction than they were before the reaction. When reacting, atoms gain, lose, or share electrons.

23. What determines an atom's chemical properties?

the atom's electron configuration

24. State the octet rule.

Eight electrons in the outer level render an atom unreactive. Helium follows this rule also, because two electrons completely fill its outer energy level.

25. Describe the general positioning of metals and nonmetals in the periodic table.

Metals are on the left and in the center and nonmetals are on the right side of the periodic table.

Chapter 7
STUDY GUIDE

7.1 SYMBOLS AND FORMULAS

Answer each of the following.

1. What is the most common source for an element's name? What are three other possible sources for the name of an element?
   a property of the element; its place of discovery, the mineral in which the element was found, a place or person to be honored by the naming

2. Complete the table by filling in the missing name or chemical symbol of each element shown.

Element	Symbol
potassium	K
gold	Au
phosphorus	P

3. a. What element does the symbol $^{40}_{20}$Ca represent? calcium
   b. What is this element's mass number? 40
   c. How many protons, electrons, and neutrons does the element contain? 20, 20, 20
   d. What would the charge be if an atom of this element lost two electrons? 2+

4. Complete the table by filling in the missing information.

Chemical formula	Names of elements in compound	Relative number of each atom
NH_3	nitrogen, hydrogen	1, 3
$C_{12}H_{22}O_{11}$	carbon, hydrogen, oxygen	12, 22, 11
ZnF_2	zinc, fluorine	1, 2
Fe_2O_3	iron, oxygen	2, 3

Write T for true or F for false. If a statement is false, change the underlined word(s) to make the statement correct by writing the correct word(s) on the blank.

__F: ions__ 5. *Atoms* are electrically charged particles.

__F: positive__ 6. In the formula of an ionic compound, the *negative* ion is written first.

__F: monatomic atom__ 7. The charge on a *diatomic molecule* is called its oxidation number.

8. Write the formulas of the compounds that will be formed when the positive ions listed in the table below combine with each negative ion listed.

	NO_3^-	S^{2-}	OH^-	Cl^-
Al^{3+}	$Al(NO_3)_3$	Al_2S_3	$Al(OH)_3$	$AlCl_3$
NH_4^+	NH_4NO_3	$(NH_4)_2S$	NH_4OH	NH_4Cl
Pb^{4+}	$Pb(NO_3)_4$	PbS_2	$Pb(OH)_4$	$PbCl_4$

Chapter 7
STUDY GUIDE

7.2 NOMENCLATURE

Match the terms listed with their descriptions. Write the correct letters on the lines provided.

__g__ 1. binary compounds
__c__ 2. hydrocarbons
__h__ 3. sulfur hexafluoride
__j__ 4. aluminum arsenate
__e__ 5. baking soda
__b__ 6. *meth-*
__a__ 7. copper(II) oxide
__i__ 8. common acids
__f__ 9. cyclopentane
__d__ 10. nitrogen

a. example of a binary compound formed from a metallic element with variable oxidation states
b. indicates one carbon atom
c. organic compounds composed solely of H and C
d. forms five different binary compounds with oxygen
e. common name of sodium hydrogen carbonate
f. has five carbon atoms linked in a ring
g. compounds that contain only two elements
h. common name of a compound containing six fluorine atoms
i. have names that do not normally follow the rules for naming compounds
j. demonstrates the rule that the name of a polyatomic ion does not end in *-ide*

11. Complete the table by filling in the missing names.

Compound formula	Name
KCl	potassium chloride
$ZnCO_3$	zinc carbonate
C_4H_{10}	butane
$FeCl_3$	iron(III) chloride
N_2O_3	nitrogen(III) oxide

Complete the sentence or answer the question.

12. What is the difference between a molecular formula and an empirical formula?
Molecular formulas are formulas for compounds that exist as molecules. Empirical formulas are the simplest formulas that indicate the ratios of atoms in compounds.

13. The formulas for ionic compounds are almost all __empirical__ formulas.

14. The molecular formula of a compound is always a whole-number multiple of the __empirical__ formula.

15. What is the empirical formula of the compound N_2O_4?
NO_2

16. How would you use symbols to designate three formula units of zinc sulfide?
3ZnS

Chapter 8
STUDY GUIDE

8.1 FACTOR-LABEL METHOD

Complete the sentence or answer the question.

1. Define the following.
   a. factor-label method <u>a problem-solving method in which unit labels are treated as factors and a check on mathematical operations is provided</u>
   b. scientific notation <u>a method or format for handling numbers in science in which all numbers are expressed as the product of a number between 1 and 10 and a whole-number power of 10</u>

2. A conversion factor is a ratio equivalent to <u>1</u>.

3. What advantage does scientific notation provide?
   <u>It removes any doubt about the number of significant digits in a measurement and makes it easier to work with very large or very small numbers.</u>

4. List the rules for handling decimal places or significant digits in the following kinds of calculations.
   a. addition and subtraction
      <u>The answer may contain only as many decimal places as the measurement that has the least number of decimal places.</u>
   b. multiplication and division
      <u>The answer may contain only as many significant digits as the measurement with the least number of significant digits.</u>

5. Complete the following conversion-factor table.

Unit given	Unit desired	Conversion factor
m	cm	100 cm/1 m
kg	g	1000 g/1 kg
nm	m	$1 \text{ m}/1 \times 10^9 \text{ nm}$
h	min	60 min/1 h
cm^3	dm^3	$1 \text{ dm}^3/1000 \text{ cm}^3$
mm	m	1 m/1000 mm
m^3	cm^3	$1 \times 10^6 \text{ cm}^3/1 \text{ m}^3$

6. Convert each of the following numbers to scientific notation.
   a. 124 <u>1.24×10^2</u>
   b. 0.000 02 <u>2×10^{-5}</u>
   c. 564.45 <u>5.6445×10^2</u>
   d. 0.1001 <u>1.001×10^{-1}</u>
   e. 2.0 <u>2.0 (or 2.0×10^0)</u>
   f. 301.03 <u>3.0103×10^2</u>

7. Enter the number of significant digits for each of the numbers below.
   a. 2.234 <u>four</u>
   b. 124.3 <u>four</u>
   c. 0.000 430 <u>three</u>
   d. 50 000 <u>one</u>
   e. 3.141 592 654 <u>ten</u>
   f. 78 456.1010 <u>nine</u>

8. Using the factor-label method, perform the following conversions. Express your answers in the correct number of significant digits.
   a. 30.0 min to h
   $$\frac{30.0 \text{ min}}{} \left| \frac{1 \text{ h}}{60 \text{ min}} \right. = 0.500 \text{ h}$$

   b. 125.1 km/s to m/min
   $$\frac{125.1 \text{ km}}{1 \text{ s}} \left| \frac{1000 \text{ m}}{1 \text{ km}} \right| \frac{60 \text{ s}}{1 \text{ min}} = 7.506 \times 10^6 \text{ m/min}$$

   c. 0.003 m to cm
   $$\frac{0.003 \text{ m}}{} \left| \frac{100 \text{ cm}}{1 \text{ m}} \right. = 0.3 \text{ cm}$$

   d. 55.0 km/h to cm/s
   $$\frac{55.0 \text{ km}}{1 \text{ h}} \left| \frac{1 \text{ h}}{60 \text{ min}} \right| \frac{1 \text{ min}}{60 \text{ s}} \left| \frac{1000 \text{ m}}{1 \text{ km}} \right| \frac{100 \text{ cm}}{1 \text{ m}} = 1.53 \times 10^3 \text{ cm/s}$$

9. Perform the following calculations. Express your answers to the correct number of significant digits and in scientific notation.

   a. add: $2.01 \times 10^3 + 4.23 \times 10^{-1}$
      $\underline{2.01 \times 10^3}$

   b. subtract: $7.2 \times 10^2 - 7.1 \times 10^3$
      $\underline{-6.4 \times 10^3}$

   c. multiply: $10.1 \times 10^2 \times 6.23 \times 10^{23}$
      $\underline{6.29 \times 10^{26}}$

   d. divide: $\dfrac{5 \times 10^{-3}}{9.0 \times 10^6}$
      $\underline{6 \times 10^{-10}}$

Chapter 8
STUDY GUIDE

8.2 FORMULA-BASED PROBLEMS

Match the term with the correct description.

a. molecular mass
b. Avogadro constant
c. molarity
d. empirical formula
e. formula mass
f. molar mass
g. percentage composition
h. molecular formula

__e__ 1. the sum of the atomic masses of all atoms in the formula unit of an ionic compound
__d__ 2. the simplest ratio of the elements in a compound
__a__ 3. the sum of all the atomic masses in a molecule
__f__ 4. the mass of 6.02×10^{23} molecules, atoms, ions, or formula units of a species
__c__ 5. the ratio between the moles of dissolved substance and the volume of solution in cubic decimeters
__h__ 6. shows the actual number of atoms in a molecule
__b__ 7. 6.02×10^{23}
__g__ 8. a statement of the relative mass each element contributes to the mass of a compound as a whole

Find the formula mass or molecular mass of the following compounds.

9. ammonia, NH_3

   1 N atom 1(14.0 u) = 14.0 u
   3 H atoms 3(1.01 u) = 3.03 u
                        _____
                        17.0 u

10. methane, CH_4

    1 C atom 1(12.0 u) = 12.0 u
    4 H atoms 4(1.01 u) = 4.04 u
                         _____
                         16.0 u

11. sodium hydrogen carbonate (baking soda), $NaHCO_3$

    1 Na atom 1(23.0 u) = 23.0 u
    1 H atom 1(1.01 u) = 1.01 u
    1 C atom 1(12.0 u) = 12.0 u
    3 O atoms 3(16.0 u) = 48.0 u
                         _____
                         84.0 u

Using the factor-label method, determine how many moles are represented by 25.0 g of each of the compounds listed in questions 9 through 11.

12. ammonia, NH_3

    $$25.0 \text{ g NH}_3 \times \frac{1 \text{ mol NH}_3}{17.0 \text{ g NH}_3} = 1.47 \text{ mol NH}_3$$

13. methane, CH_4

    $$25.0 \text{ g CH}_4 \times \frac{1 \text{ mol CH}_4}{16.0 \text{ g CH}_4} = 1.56 \text{ mol CH}_4$$

14. sodium hydrogen carbonate, $NaHCO_3$

    $$25.0 \text{ g NaHCO}_3 \times \frac{1 \text{ mol NaHCO}_3}{84.0 \text{ g NaHCO}_3} = 0.298 \text{ mol NaHCO}_3$$

Using the factor-label method, determine the mass, in grams, of 2.50×10^{24} molecules or formula units of each of the compounds in questions 9 through 11.

15. ammonia, NH_3

    $$2.50 \times 10^{24} \text{ molecules} \times \frac{1 \text{ mol NH}_3}{6.02 \times 10^{23} \text{ molecules}} \times \frac{17.0 \text{ g NH}_3}{1 \text{ mol NH}_3} = 70.6 \text{ g NH}_3$$

16. methane, CH_4

    $$2.50 \times 10^{24} \text{ molecules} \times \frac{1 \text{ mol CH}_4}{6.02 \times 10^{23} \text{ molecules}} \times \frac{16.0 \text{ g CH}_4}{1 \text{ mol CH}_4} = 66.4 \text{ g CH}_4$$

17. sodium hydrogen carbonate, $NaHCO_3$

    $$2.50 \times 10^{24} \text{ formula units} \times \frac{1 \text{ mol NaHCO}_3}{6.02 \times 10^{23} \text{ formula units}} \times \frac{84.0 \text{ g NaHCO}_3}{1 \text{ mol NaHCO}_3} = 349 \text{ g NaHCO}_3$$

18. Complete the following table by filling in the correct values.

Concentration	Moles of solute	Volume of solution
10.0M	30.0 mol	3.00 dm³
0.062M	0.651 mol	10.5 dm³
1.1M	6.3 mol	5.9 dm³
6.0M	6.0 mol	1.0 dm³
0.111M	0.0555 mol	0.500 dm³

Describe the preparation of each of the following solutions.

19. 1.00 dm³ of 6.00M NaCl

    $$1.00 \text{ dm}^3 \times \frac{6.00 \text{ mol NaCl}}{1 \text{ dm}^3} \times \frac{58.5 \text{ g NaCl}}{1 \text{ mol NaCl}} = 351 \text{ g NaCl}$$

    Dissolve 351 g NaCl in sufficient water to make 1.00 dm³ of solution.

20. 2.50 dm³ of 0.100M $BaCl_2$

    $$2.50 \text{ dm}^3 \times \frac{0.100 \text{ mol BaCl}_2}{1 \text{ dm}^3} \times \frac{208 \text{ g BaCl}_2}{1 \text{ mol BaCl}_2} = 52.0 \text{ g BaCl}_2$$

    Dissolve 52.0 g $BaCl_2$ in sufficient water to make 2.50 dm³ of solution.

Find the percentage composition of the following compounds.

21. carbon dioxide, CO_2

    1 C atom $1(12.0\ u) = 12.0\ u$
    2 O atoms $2(16.0\ u) = 32.0\ u$
    $44.0\ u$

    percentage of C = $\dfrac{mass\ 1C}{mass\ CO_2} \times 100\% = \dfrac{12.0\ u}{44.0\ u} \times 100 = 27.3\%$

    percentage of O = $\dfrac{mass\ 2O}{mass\ CO_2} \times 100\% = \dfrac{32.0\ u}{44.0\ u} \times 100 = 72.7\%$

22. sulfuric acid, H_2SO_4

    2 H atoms $2(1.01\ u) = 2.02\ u$
    1 S atom $1(32.1\ u) = 32.1\ u$
    4 O atoms $4(16.0\ u) = 64.0\ u$
    $98.1\ u$

    percentage of H = $\dfrac{mass\ 2H}{mass\ H_2SO_4} \times 100\% = \dfrac{2.02\ u}{98.1\ u} \times 100 = 2.06\%$

    percentage of S = $\dfrac{mass\ 1S}{mass\ H_2SO_4} \times 100\% = \dfrac{32.1\ u}{98.1\ u} \times 100 = 32.7\%$

    percentage of O = $\dfrac{mass\ 4O}{mass\ H_2SO_4} \times 100\% = \dfrac{64.0\ u}{98.1\ u} \times 100 = 65.2\%$

23. hydrogen peroxide, H_2O_2

    2 H atoms $2(1.01\ u) = 2.02\ u$
    2 O atoms $2(16.0\ u) = 32.0\ u$
    $34.0\ u$

    percentage of H = $\dfrac{mass\ 2H}{mass\ H_2O_2} \times 100\% = \dfrac{2.02\ u}{34.0\ u} \times 100\% = 5.94\%$

    percentage of O = $\dfrac{mass\ 2O}{mass\ H_2O_2} \times 100\% = \dfrac{32.0\ u}{34.0\ u} \times 100\% = 94.1\%$

Calculate the empirical formulas of the following.

24. a compound that is 45.9% K, 16.5% N, and 37.6% O

    $45.9\ \text{g K} \left| \dfrac{1\ \text{mol K}}{39.1\ \text{g K}} \right. = 1.17\ \text{mol K}$

    $16.5\ \text{g N} \left| \dfrac{1\ \text{mol N}}{14.0\ \text{g N}} \right. = 1.18\ \text{mol N}$

    $37.6\ \text{g O} \left| \dfrac{1\ \text{mol O}}{16.0\ \text{g O}} \right. = 2.35\ \text{mol O}$

    $1.17 : 1.18 : 2.35 :: 1 : 1 : 2;\ KNO_2$

25. a compound, 5.00 g of which contains 4.28 g C and 0.720 g H

    $4.28\ \text{g C} \left| \dfrac{1\ \text{mol C}}{12.0\ \text{g C}} \right. = 0.357\ \text{mol C}$

    $0.720\ \text{g H} \left| \dfrac{1\ \text{mol H}}{1.01\ \text{g H}} \right. = 0.713\ \text{mol H}$

    $0.357 : 0.713 :: 1 : 2;\ CH_2$

26. a compound, a 100.0-g sample of which contains 11.2 g H and 88.8 g O

    $11.2\ \text{g H} \left| \dfrac{1\ \text{mol H}}{1.01\ \text{g H}} \right. = 11.1\ \text{mol H}$

    $88.8\ \text{g O} \left| \dfrac{1\ \text{mol O}}{16.0\ \text{g O}} \right. = 5.55\ \text{mol O}$

    $11.1 : 5.55 :: 2 : 1;\ H_2O$

Calculate the molecular formulas of the compounds listed below, given the empirical formula and formula mass of each.

27. empirical formula = CH, formula mass = 78.1 u.

    The formula mass of empirical unit CH is 13.0 u.

    $\dfrac{78.1\ u}{13.0\ u} = 6.01 \cong 6.$

    The molecular formula is C_6H_6.

28. empirical formula = NH_2, formula mass = 32.1 u.

    The formula mass of empirical unit NH_2 is 16.0 u.

    $\dfrac{32.1\ u}{16.0\ u} = 2.01 \cong 2.$

    The molecular formula is N_2H_4.

Chapter 9
STUDY GUIDE

9.1 CHEMICAL EQUATIONS

Complete each sentence.

1. The process by which one or more substances are changed into one or more different substances is called a(n) ___chemical reaction___.

2. The starting substances in a chemical reaction are called ___reactants___.

3. The symbol written after a formula to indicate a solid is ___(cr)___.

4. In a balanced chemical equation, the number of ___atoms___ of any given kind on the left side is equal to the number on the right side.

Match each reaction type with the correct description.

- b 5. The positive and negative portions of two compounds are interchanged.
- a 6. A compound burns, reacting with oxygen.
- c 7. Two or more substances combine to form one new substance.
- e 8. A substance breaks down to form simpler substances when energy is supplied.
- d 9. One element replaces another in a compound.

 a. combustion
 b. double displacement
 c. synthesis
 d. single displacement
 e. decomposition

Write T for true or F for false. If a statement is false, replace the underlined word or phrase with one that will make the statement true, and write your correction on the blank provided.

- T   10. The symbol (aq) after a formula indicates that the substance is ___dissolved in water___.
- F; coefficient   11. A number written to the left of a formula in an equation is called a ___subscript___.
- F; Carbon dioxide   12. ___Carbon___ and water are typical products in a combustion reaction.
- T   13. The substances generally written on the right in chemical equations are called ___products___.
- F; an arrow   14. In a chemical equation, the symbol that stands for *yields* is ___a plus sign___.
- F; single displacement   15. The general form of a ___double displacement___ reaction is: element + compound → element + compound.
- F; is not   16. In balancing a chemical equation, it ___is___ permissible to change subscripts.

9.2 STOICHIOMETRY

Complete each sentence.

1. The branch of chemistry that deals with the amounts of substances involved in chemical reactions is called ___stoichiometry___.

2. In solving a mass-mass problem, one must convert the number of grams of the given substance to ___moles___.

3. The actual amount of product divided by the theoretical amount and multiplied by 100 is called the ___percentage yield___.

4. A(n) ___mass-energy___ problem is one in which the amount of heat absorbed or released during a reaction is calculated from information on the number of grams.

Write T for true or F for false. If a statement is false, replace the underlined word or phrase with one that will make the statement true, and write your correction on the blank provided.

- F; moles   5. The coefficients in a balanced chemical equation show the correct ratio of ___masses___.
- F; necessary   6. In solving a mass-mass problem, it is ___not necessary___ to work with a balanced chemical equation.
- T   7. In a reaction with a low percentage yield, the amount of product produced, compared with the amount that theoretically could have been produced, is ___small___.
- T   8. The heat of reaction is represented by the letter *q*.
- F; negative   9. The heat of reaction for a system giving off energy has a ___positive___ value.
- T   10. To convert grams to moles, the number of grams term on the left represents a factor that has the units mol/g.
- T   11. A chemical equation that shows an energy term on the left represents a reaction that ___absorbs___ energy.

Chapter 10 STUDY GUIDE

10.1 PERIODIC TRENDS

Write a definition for each of the following terms.

1. atomic radius — the radius of an atom without regard to surrounding atoms

2. noble gas configuration — stable electron configurations with filled outer energy levels like those of noble gases

Answer each question in complete sentences.

3. Why does an atom of sodium have a larger atomic radius than an atom of chlorine has, even though sodium has fewer electrons?

   The outer electrons of both atoms are in the third energy level. Chlorine has six more protons in its nucleus than does sodium, exerting a greater attractive force on its electrons, and therefore pulling them in closer.

4. When dissolved in water, NaCl will carry an electric current, but when the NaCl is in solid form, it will not do so. Explain this fact.

   When dissolved, charged Na^+ and Cl^- ions are free to flow and carry current. In solid NaCl, the ions are bound in the crystal structure and cannot flow.

5. If iron has an electron configuration of $1s^2 2s^2 2p^6 3s^2 3p^6 4s^2 3d^6$, how does it form ions with oxidation numbers of 2+ and 3+?

   Iron will first lose its outer-level $4s^2$ electrons (giving it a 2+ oxidation number), but because a half-full d sublevel would be especially stable, the $3d$ level can readily lose one electron to achieve this stable $3d^5$ state, creating a 3+ oxidation number.

6. Predict the most likely oxidation number for atoms that have the following electron configurations.

   a. $1s^2 2s^2 2p^6 3s^2 3p^6 4s^2 3d^{10}$ — 2+
   b. $1s^2 2s^2 2p^6 3s^2 3p^5$ — 1−
   c. $1s^2 2s^2 2p^6 3s^2 3p^1$ — 3+
   d. $1s^2 2s^2 2p^6 3s^2 3p^6 4s^2 3d^{10} 4p^6 5s^1 4d^{10}$ — 1+

10.2 REACTION TENDENCIES

Write a definition for each of the following terms.

1. ionization energy — the energy required to remove an electron from an atom

2. first ionization energy — the amount of energy needed to remove the most loosely held electron from an atom

3. shielding effect — a decrease in the attractive force between outer electrons and the nucleus, caused by the presence of inner electrons

4. electron affinity — the attraction of an atom for an additional electron

Answer the following question in a complete sentence.

5. State the four factors that affect ionization energy, and briefly describe the effect of each.

   nuclear charge: the larger the charge is, the greater the ionization energy is; shielding effect: the greater the shielding effect is, the less the ionization energy is; radius: the greater the distance is from the nucleus to the outer level of an atom, the less the ionization energy is; full or half-full sublevel: an electron from such a sublevel requires additional energy to be removed

Complete each of the following statements.

6. The first ionization energy tends to increase as atomic number increases in any horizontal row, or period.

7. A column, or group, will show a decrease in first ionization energy as atomic number increases.

8. Ionization energy is typically measured in units called kilojoules per mole (kJ/mol).

9. Metals have a low first ionization energy, and nonmetals have a high first ionization energy.

10. Nonmetals have high electron affinities. Metals have low electron affinities. Moving down a column, the values of the elements' electron affinities tend to decrease.

Chapter 11 STUDY GUIDE

11.1 Hydrogen and Main Group Metals

Write T for true or F for false. If a statement is false, replace the underlined word or phrase with one that will make the statement true, and write your correction on the blank provided.

__T__ 1. The shielding effect causes large atoms to lose their electrons <u>more</u> readily than they might otherwise.

__F; hydrogen__ 2. A <u>hydride</u> ion is a bare proton.

__T__ 3. A catalyst is a substance that <u>speeds up</u> a reaction.

__T__ 4. By gaining an electron, hydrogen attains the stable electron configuration of <u>helium</u>.

__T__ 5. As atoms of the alkali metals <u>increase</u> in size, they lose their outermost electron more easily.

__F; sodium hydrogen carbonate__ 6. Baking soda is another name for <u>sodium carbonate</u>.

__T__ 7. The thallium ion and the ammonium ion each have a charge of <u>1+</u>.

__F; lithium__ 8. The alkali metal that is insoluble in most other alkali metals is <u>cesium</u>.

__F; alkaline earth metals__ 9. The <u>alkali metals</u> make up Group 2 of the periodic table.

__T__ 10. Lime is a <u>calcium</u> compound.

__F; Aluminum__ 11. <u>Magnesium</u> is the most plentiful metal in Earth's crust.

__T__ 12. Alkaline earth metals tend to form <u>2+</u> ions.

Match each element with the correct description.

a. sodium   f. calcium
b. francium   g. aluminum
c. hydrogen   h. beryllium
d. potassium   i. lithium
e. magnesium   j. radium

__b__ 13. the most active metal
__h__ 14. is used in making nonsparking tools
__a__ 15. forms a silicate used as a catalyst and in soapmaking
__d__ 16. found in large amounts in muscle and nerve tissue
__c__ 17. can form both positive and negative ions
__f__ 18. its ions are a major constituent of bone and affect release and absorption of hormones
__i__ 19. the largest alkaline earth metal atom
__g__ 20. has three outermost electrons
__e__ 21. is part of a chlorophyll molecule
__j__ 22. alkali metal that reacts most vigorously with nitrogen and that burns in air to form an oxide, rather than a peroxide

11.2 Nonmetals

Complete each sentence.

1. When most of the elements in Group 14 (IVA) react, they tend to __share__ electrons.
2. Different forms of the same element are called __allotropes__.
3. Dry ice is the name for solid __carbon dioxide__.
4. The Haber process produces __ammonia__.
5. The transfer of genetic information from generation to generation involves a nitrogen-phosphorus organic compound called __DNA (deoxyribonucleic acid)__.
6. The oxide ion has a charge of __2−__.
7. The form of oxygen that has the formula O_3 is called __ozone__.
8. When SO_3 is dissolved in water, the compound called __sulfuric acid__ is formed.
9. Ions made up of long chains of sulfur atoms attached to the S^{2-} ion are called __polysulfide__ ions.
10. Chlorine and bromine are members of the __halogen__ family.
11. The least reactive elements make up a family called the __noble gases__.
12. A mixture of metals is called a(n) __alloy__.
13. A mixture of copper and tin is called __bronze__.
14. When the members of Group 17 (VIIA) react, they usually __gain__ electrons.
15. Going down Group 17 (VIIA) of the periodic table, there is a(n) __decrease__ in the activity of the elements.

Match each element with its description. Some choices may serve as answers more than once.

a. argon   g. selenium
b. oxygen   h. silicon
c. xenon   i. phosphorus
d. lead   j. nitrogen
e. fluorine   k. helium
f. carbon

__e__ 16. the most reactive nonmetallic element
__b__ 17. the most plentiful element in Earth's crust
__f__ 18. its allotropes are graphite and diamond
__h__ 19. the second most plentiful element in Earth's crust
__d__ 20. a distinctly metallic element in Group 14 (IVA)
__a__ 21. used in light bulbs to protect the filament
__f__ 22. a constituent of all organic compounds
__j__ 23. makes up most of Earth's atmosphere

Chapter 11
STUDY GUIDE

11.3 TRANSITION METALS

Complete each sentence.

1. Steel is an alloy of   iron .
2. The highest-energy electrons of transition metals are in the   d   sublevel.
3. There is a total of   10   columns, or groups, of transition metals in the periodic table.
4. The CrO_4^{2-} ion is called the   chromate   ion.
5. In most reactions, chromium loses   three   of its electrons.
6. Iron is galvanized by being dipped into molten   zinc .
7. Rubies and emeralds get their color from   chromium   impurities.
8.   Brass   is an alloy of copper and zinc.
9. The highest-energy electrons of the inner transition elements are in the   f   sublevel.
10. The inner transition elements of Period 6 are called   lanthanoids .
11. The inner transition elements of Period 7 are called   actinoids .
12. The most stable ion for lanthanoids has a charge of   3+ .
13. The lanthanoid element   neodymium   forms alloys with unusual conductivity and magnetic properties.
14.   Curium   is a highly toxic actinoid.
15. A transition metal that is found in many proteins in biological systems is   zinc .

Match each element with its use described below.

a. silver
b. molybdenum
c. cobalt
d. zinc
e. manganese
f. neodymium
g. chromium
h. curium
i. titanium
j. osmium

__i__ 16. forms a dioxide used as a white paint pigment
__b__ 17. is used in spark plugs
__d__ 18. is needed in the diet for proper functioning of the pancreas
__j__ 19. is used to harden pen points
__e__ 20. forms a dioxide used in batteries
__a__ 21. is used in coinage
__g__ 22. is used to plate steel to protect it from corrosion
__h__ 23. may be used in the future as the energy source in satellite nuclear generators
__c__ 24. has a radioactive isotope used in cancer treatment
__f__ 25. forms an oxide used in glass filters and lasers

__c__ 24. the first noble gas to produce a compound
__h__ 25. a semiconductor used in transistors and computer chips
__d__ 26. produces solder when mixed with tin
__i__ 27. in its white form, ignites spontaneously in air
__k__ 28. was first discovered somewhere other than on Earth
__g__ 29. a metalloid member of Group 16 (VIA)
__f__ 30. the only element to exhibit catenation to a great extent

Write T for true or F for false. If a statement is false, replace the underlined word or phrase with one that will make the statement true, and write your correction on the blank provided.

__F; inorganic__ 31. Carbonic acid and cyanides are examples of <u>organic</u> compounds.
__T__ 32. <u>Tin</u> is an example of a metal that has been known since prehistoric times.
__F; nitrogen__ 33. Some bacteria convert atmospheric <u>oxygen</u> to compounds that can be used readily by plants to make amino acids.
__F; P_4__ 34. Elemental phosphorus occurs as <u>P_2</u> molecules.
__T__ 35. A Group 15 (VA) element that occurs in all oxidation states from 3− to 5+ is <u>nitrogen</u>.
__T__ 36. <u>Argon</u> is an example of an element that has never been made to form compounds.
__F; noble gas__ 37. Elements in the <u>carbon</u> family have complete outer energy levels.

Chapter 12
STUDY GUIDE

12.1 BOND FORMATION

1. Define electronegativity.
   Electronegativity is the relative tendency of an atom to attract electrons to itself when it is bonded to another atom.

2. Electron affinity and electronegativity are both measures of an atom's attraction for electrons. What is the difference between them?
   Electron affinity is a property of an isolated atom; electronegativity is a property of the atom when it is bonded to another atom.

3. How does electronegativity vary as the atomic number of an element increases within the same period of the periodic table?
   The electronegativity increases.

4. How is the strength of a bond between two elements in a molecule related to their electronegativities?
   The greater the difference in electronegativity, the greater the bond strength.

5. What is the difference between an ionic and a covalent bond?
   In an ionic bond, electrons are transferred from one atom to another; that is, one atom gains, and the other loses, electrons. In a covalent bond, two atoms share electrons.

6. How is the character of a bond (ionic or covalent) between two elements related to their electronegativities?
   The bond character depends on the difference in their electronegativities. If the difference is less than 1.67, the bond is more covalent than ionic; if the difference is greater than 1.67, the bond is more ionic than covalent.

7. Referring to Table 12.1, Electronegativities, in your text, arrange the following compounds in order of increasing ionic character of their bonds: LiF, LiBr, KCl, KI.
   KI, LiBr, KCl, LiF

8. Referring to Tables 12.1 and 12.3 in your text, classify each of the following bonds as either ionic (I) or covalent (C):

   __I__ a. Al–O __C__ f. N–O
   __C__ b. Al–S __C__ g. Na–S
   __C__ c. Bi–Cl __C__ h. P–O
   __I__ d. Bi–O __C__ i. S–O
   __C__ e. C–Cl __C__ j. Tl–Br

9. What force holds the two ions together in an ionic bond?
   The electrostatic force due to their opposite charges holds the ions together.

10. What is the meaning of the oxidation number of an element that forms an ionic bond?
    The oxidation number of an element in an ionic bond is numerically equal to the number of electrons it has gained or lost and is negative if it has gained electrons and positive if it has lost electrons.

11. What is a molecule?
    A molecule is the particle formed when two or more atoms are bonded to one another covalently.

12. What causes the bond lengths and bond angles of a molecule to vary?
    Bond lengths and angles vary as the atoms in a molecule vibrate.

13. a. Why does a molecular compound absorb specific frequencies of infrared radiation?
    The molecule of a compound absorbs those frequencies of infrared radiation that correspond to its natural frequencies of vibration.

    b. How can the absorption of this radiation be used to identify a compound?
    Every molecular compound has a different set of natural frequencies. It can therefore be recognized by the particular frequencies of infrared radiation that it absorbs.

14. Why are the bonding electrons in metallic bonding said to be delocalized?
    In metallic bonding, the bonding electrons are not associated with any particular atoms and therefore are not "localized" in the substance.

15. What is one factor that determines the hardness of a metallic element?
    The hardness of a metal tends to increase with the number of outer electrons per atom that are available for bonding.

16. Indicate whether each property listed below is characteristic of ionic (I), covalent (C), or metallic (M) bonding. More than one letter may be used for each answer.
    __M__ a. Shape of solid can be changed by pounding.
    __I,C__ b. Not electrically conducting in solid phase
    __M__ c. Electrically conducting in all phases
    __I,M__ d. High melting points

17. a. How are a molecule and a polyatomic ion alike?
    The atoms in both are bonded covalently.

    b. How are they different?
    A molecule is electrically neutral, while a polyatomic ion has an electric charge.

Chapter 12
STUDY GUIDE

12.2 PARTICLE SIZES

1. a. Which is larger, a metallic atom or its positive ion?
   A metallic atom is larger than its positive ion.
   b. Which is larger, a nonmetallic atom or its negative ion?
   The negative ion of a nonmetallic atom is larger than the atom.

2. In an ionic crystal, how can the internuclear distance between two ions be calculated?
   The internuclear distance between two ions is the sum of their ionic radii.

3. In a molecule, how can the approximate bond length between two atoms be calculated?
   The approximate bond length between two atoms in a molecule is the sum of their covalent radii.

4. Use Table 12.5 in your textbook to calculate the expected bond length (in picometers) of the following covalent bonds:

   __184__ a. P–O
   __144__ b. N–O
   __140__ c. N–N

5. Arrange the atomic radius, covalent radius, and ionic radius of a nonmetallic atom in the expected order of increasing size.
   covalent radius, atomic radius, ionic radius

6. Under what circumstances might the actual order of these sizes change?
   When an atom is bonded to two or more other atoms, its electron cloud may be so distorted that its covalent radius becomes larger than its atomic radius.

Chapter 13
STUDY GUIDE

13.1 BONDS IN SPACE

Complete the sentence or answer the question.

1. Pairs of electrons that bond two atoms in a molecule are called __shared pairs__.

2. Why do electron pairs in the outer levels of atoms in a molecule spread apart as far as possible?
   to minimize the repulsive forces between their charge clouds

3. Because each of the bond angles of methane equals 109.5°, its molecular shape is a perfect __tetrahedron__.

4. Complete the following diagrams to show the electron content of the outer orbitals of (a) an unbonded carbon atom, and (b) a hybridized carbon atom.

   Unbonded carbon atom   Hybridized carbon atom
   Outer orbitals   Outer orbitals

   $2s$ $2p$ $2sp^3$
   (↑↓) (↑)(↑)( )   (↑)(↑)(↑)(↑)

5. How do hybridized orbitals in an atom compare with one another?
   Hybridized orbitals in any atom are exactly alike, except for orientation in space.

6. When an sp^3 orbital of one carbon atom overlaps that of another carbon atom, how many electrons do the two atoms share?
   2

7. How many hybrid orbitals of the following types can an atom have?
   a. sp^3 __4__
   b. sp^2 __3__
   c. sp __2__

8. In a molecule of ethylene, C_2H_4, how many sigma bonds and how many pi bonds are present?
   3 sigma bonds and 2 pi bonds

9. In a trigonal planar structure such as the CH_2O molecule, with one double bond and two single bonds, how does the double bond affect the angle between the two single bonds?
   The double bond causes the angle between the single bonds to become less than 120°.

Chapter 13
STUDY GUIDE

13.2 MOLECULAR ARRANGEMENTS

Complete the sentence or answer the question.

1. In saturated organic compounds, all the bonds between carbon atoms are __single__.

2. The existence of two or more substances with the same molecular formula but different structural formulas is known as __isomerism__.

3. What is the difference between structural isomerism and positional isomerism?
   In structural isomers, only the carbon chain is altered. In positional isomerism, the carbon chain remains the same, but the position of something, such as a double bond, is changed.

4. Compounds with the same atoms bonded in the same order but with a different arrangement of atoms around a double bond are examples of __geometric isomers__.

5. a. What type of hybridization would be expected in an atom with three outer electrons? sp^2
   b. What would be the shape and bond angles of the molecule it forms? trigonal planar with 120° bond angles

Match each suffix with its correct description of the bonding of the carbon atoms in a hydrocarbon.

a. saturated compounds (single bonds only)
b. unsaturated compounds with double bonds
c. unsaturated compounds with triple bonds

__c__ 6. -yne
__a__ 7. -ane
__b__ 8. -ene

Match each of the following compounds with the correct description of its molecule.

a. boron trichloride (BCl_3)
b. acetylene (ethyne) (C_2H_2)
c. ozone (O_3)
d. oxygen difluoride (OF_2)
e. carbon tetrachloride (CCl_4)
f. trimethylarsine ($(CH_3)_3As$)

__b__ 9. linear, with 180° bond angles
__d__ 10. bent, with an exceptionally small bond angle resulting from two unshared pairs of electrons
__c__ 11. bent, with a small bond angle resulting from one unshared pair of electrons
__e__ 12. tetrahedral, with the expected bond angles of 109.5°
__f__ 13. trigonal pyramidal, with an unusually small bond angle (96°)
__a__ 14. trigonal planar, with the expected bond angles of 120°

Write T for true or F for false. If a statement is false, replace the underlined word or phrase with one that will make the statement true, and write your correction on the blank provided.

__T__ 10. A shared pair of electrons is formed when an orbital of one atom overlaps an orbital of another atom.

__F; pi__ 11. Sigma bonds are always formed by the sideways or parallel overlap of unhybridized *p* orbitals.

__F; sigma__ 12. In single, double, and triple bonds between carbon atoms, one of the bonds is always a pi bond.

__T__ 13. When two carbon atoms are joined by more than one bond, the additional bonds are always pi bonds.

__F; farther from__ 14. Pi bonds break more easily than do sigma bonds, because the electrons forming pi bonds are closer to the nuclei.

For each of the following pairs of isomers, describe the difference in structure and identify the type of isomerism illustrated.

15.

```
   H H H              H OH H
   | | |              |  |  |
 H-C-C-C-OH         H-C--C--C-H
   | | |              |  |  |
   H H H              H  H  H

   1-propanol          2-propanol
```

Structural difference: The position of the OH group changed from the first carbon to the second.

Type of isomerism: positional isomerism

16.

```
   H H                H      H
   | |                |      |
 H-C-C-OH           H-C--O--C-H
   | |                |      |
   H H                H      H

   ethanol            methoxymethane
```

Structural difference: The oxygen (O) atom is bonded in two different ways.

Type of isomerism: functional isomerism

Complete the sentence or answer the question.

17. a. In the compound named ethene, what does the ending -ene indicate about its structure?

The ending -ene indicates that the molecule has a pair of carbon atoms with a double bond between them.

b. What does the stem eth- indicate?

The stem eth- indicates that the molecule has 2 carbon atoms in the carbon chain being named.

18. a. Draw the structural formula for CH≡C—CH$_2$—CH$_3$.

```
          H H
          | |
 H-C≡C-C-C-H
          | |
          H H
```

b. Give the name of this compound, and explain the meaning of its stem and ending.

The name of the compound is butyne. The stem but- indicates 4 carbons in the main carbon chain, and the ending -yne indicates a triple bond.

19. Under each of the following simplified structural diagrams, write the name of the compound represented.

cyclobutene   cyclopentane   cyclohexene   cyclooctane

20. Draw the structural formula for the positional isomer of the compound in question 18.

```
       H       H
       |       |
   H-C-C≡C-C-H
       |       |
       H       H
```

Chapter 14
STUDY GUIDE

14.1 MOLECULAR ATTRACTION

Complete the sentence or answer the question.

1. Define the following terms.
   a. polar covalent bond   a covalent bond between the atoms of two different elements in which one of the atoms attracts the shared pair of electrons more strongly than does the other
   b. dipole   an atom or molecule that has both a positive and negative pole
   c. dipole moment   a measure of the strength of the dipole

2. Are the polar bonds in a polar molecule arranged symmetrically or asymmetrically?
   asymmetrically

3. What does the dipole moment of a molecule tell us about its intermolecular forces?
   the higher the dipole moment, the stronger the intermolecular forces

4. Indicate whether the following molecules are polar or nonpolar.
   a. BeF_2 __nonpolar__
   b. H_2O __polar__
   c. $CHCl_3$ __polar__
   d. CCl_4 __nonpolar__

Write the letter of the term that matches the correct description.

a. van der Waals forces
b. intramolecular forces
c. intermolecular forces
d. dipole-dipole forces
e. dipole-induced dipole forces
f. temporary dipole
g. dispersion forces
h. induced dipole

__f__ 5. a nonpolar molecule in which the charge distribution is briefly asymmetrical

__h__ 6. a nonpolar molecule that has its electron cloud distorted by an approaching dipole and is thus transformed into a dipole

__a__ 7. weak forces involving the attraction of the electrons of one atom for the protons of another

__d__ 8. attractive forces between two molecules of the same or different substances that are both permanent dipoles

__g__ 9. forces generated by temporary dipoles

__c__ 10. forces between molecules

__e__ 11. attractive forces between dipoles and nonpolar molecules

__b__ 12. forces within a molecule

Write T for true and F for false. If a statement is false, replace the underlined word or phrase with one that will make the statement true, and write your correction on the blank provided.

__T__ 13. <u>Dispersion</u> forces are the only attractive forces that attract between nonpolar molecules.

__F; can__ 14. Molecules <u>cannot</u> exhibit both dipole and dispersion interactions.

__F; low-boiling liquids__ 15. Substances composed of nonpolar molecules are generally gases at room temperature or <u>high-boiling liquids</u>.

__T__ 16. Substances composed of polar molecules generally have <u>higher</u> boiling points than do nonpolar compounds.

__T__ 17. A nonpolar molecule <u>can</u> have polar bonds.

__F; polar__ 18. Water is a <u>nonpolar</u> molecule that has polar bonds.

Chapter 14
STUDY GUIDE

14.2 COORDINATION CHEMISTRY

Complete the sentence or answer the question.

1. What is a complex ion?
   a cluster of ions formed by polar molecules or negative ions around a central positive ion

2. Name two uses for complex ions.
   Sample answers include: in analytical chemistry or as catalysts in industry.

3. A ligand is either a polar molecule or a negative ion that is attached to a central positive ion in a complex ion.

4. The coordination number of a complex ion is the number of points of attachment of ligands surrounding the central positive ion.

5. Why is the oxalate ion in a complex ion called a didentate ligand?
   It attaches to the central ion at two points.

6. What type of complex is formed by three didentate ligands? octahedral

7. The molecules of a coordination compound are complexes with a net charge of zero .

Write T for true and F for false. If a statement is false, replace the underlined word or phrase with one that will make the statement true, and write your correction on the blank provided.

F; polar   8. In a complex ion, <u>nonpolar</u> molecules or negative ions are attached to a central positive ion.

F; negative   9. Ligands can be either molecules or <u>positive</u> ions.

T   10. The most common ligand is <u>water</u>.

T   11. Complex ions with coordination number <u>two</u> are always linear.

F; the same atom   12. In a coordinate covalent bond the electrons in the shared pair come from <u>different atoms</u>.

F; covalent   13. Although the bonds of most complex ions have characteristics of both covalent and ionic bonding types, the <u>ionic</u> character dominates.

14. Complete the following table.

Coordination number	Shape of complex
2	linear
4	tetrahedral
	square planar
6	octahedral

Chapter 14
STUDY GUIDE

14.3 CHROMATOGRAPHY

Complete the sentence or answer the question.

1. Define the following terms.
   a. fractionation any process of separating substances (parts) from a mixture (whole)
   b. chromatography a process of fractionation in which a mobile phase containing a mixture of substances is allowed to pass over a stationary phase that has attraction for polar materials

2. Which phase in chromatography consists of the mixture to be separated being dissolved in a fluid (liquid or gas)? mobile phase

3. Label the following parts of the gas chromatography system in the figure.
   a. inert carrier gas
   b. stationary phase
   c. mobile phase
   d. volatile mixture to be separated

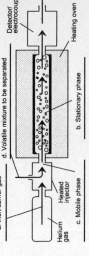

Write T for true or F for false. If a statement is false, replace the underlined word or phrase with one that will make the statement true, and write your correction on the blank provided.

T   4. Chromatography is a method of separating a mixture of substances into components based on differences in <u>polarity</u> of the components.

T   5. In chromatography, the fastest migrating substance will be the one with the <u>least</u> attraction for the stationary phase.

F; high performance liquid   6. <u>Paper</u> chromatography is used when it is necessary to speed up the separation process.

F; ion   7. An <u>ion exchange</u> resin is used as the stationary phase of <u>thin layer</u> chromatography.

Write the letter of the type of chromatographic technique that matches the description.

a. column chromatography
b. high performance liquid chromatography
c. ion chromatography
d. paper chromatography
e. thin layer chromatography
f. gas chromatography

e   8. combines some of the techniques of both column and paper chromatography and is used frequently in separating biological materials

f   9. is a chromatographic technique for the analysis of volatile liquids and mixtures of gases

b   10. is designed to overcome the speed limitations of conventional column chromatography

d   11. is a simple and fast chromatographic technique, but one in which quantitative determinations are difficult because of its extremely small scale

a   12. is used for vitamins, proteins, and hormone separations not easily made by other methods

Chapter 15
STUDY GUIDE

15.1 PRESSURE

Write T for true or F for false. Is a statement is false, replace the underlined word or phrase with one that will make the statement true, and write your correction in the blank.

__T__ 1. All matter is composed of <u>small particles</u>.

__F; in constant motion__ 2. Particles that make up matter are always <u>stationary</u>.

__T__ 3. Collisions between particles of matter are perfectly elastic because there is <u>no change</u> in the total kinetic energy of the particles before and after the collision.

__F; billion__ 4. Under ordinary conditions, each molecule of a gas undergoes a few <u>million</u> collisions each second.

__F; a closed-arm__ 5. A barometer is <u>an open-arm</u> manometer used to measure atmospheric pressure.

__F; 101.325 kilopascals (kPa)__ 6. Standard atmospheric pressure is <u>one pascal</u>.

Complete the statement or answer the question.

7. What is one pascal?
   One pascal is a pressure of one newton per square meter.

8. In using any manometer, what measurement must you make?
   You must measure the difference in liquid level between the two arms.

9. What physical property of the liquid in a manometer must you know?
   You must know the density of the liquid.

10. Why is the measurement of gas pressure with a closed-arm manometer independent of the atmospheric pressure?
    There is a vacuum above the liquid in the closed arm.

11. Standard atmospheric pressure will support a column of mercury __760__ mm high.

12. One kilopascal equals __7.501__ mm Hg.

13. A closed-arm mercury manometer is shown being used to measure the pressure of a gas in a closed container. What is the pressure of the gas in kilopascals?

    $\dfrac{500.0 \text{ mm Hg}}{1} \times \dfrac{1 \text{ kPa}}{7.501 \text{ mm Hg}} = 66.66 \text{ kPa}$

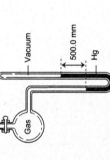

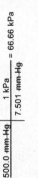

Chapter 15
STUDY GUIDE

15.2 MOTION AND PHYSICAL STATES

Answer the following questions.

1. Complete the following table comparing the three common states of matter.

State of matter	Shape	Volume
solid	definite	definite
liquid	assumes the shape of its container	definite
gas	assumes the shape of its container	fills its container completely

2. What two factors determine the average speed of the particles in a gas?
   The average speed of the particles depends on the temperature of the gas and the mass of the particles.

3. Define *absolute zero*.
   Absolute zero is the temperature at which all molecular motion theoretically ceases.

4. In which direction does energy flow between two objects at different temperatures?
   Energy always flows from a warmer object to a cooler object.

5. What is the SI unit of temperature?
   The Kelvin is the SI unit of temperature.

6. What is absolute zero on the Celsius temperature scale?
   −273.15°C (−273°C)

7. What equation can be used to convert Celsius temperatures (°C) to Kelvin temperatures (K)?
   K = °C + 273

8. Convert the following temperatures to the Kelvin scale.
   a. 15°C __288 K__
   b. −100°C __173 K__
   c. 0°C __273 K__

9. Convert the following temperatures to the Celsius scale.
   a. 0 K __−273°C__
   b. 100 K __−173°C__
   c. 722 K __449°C__

10. What is the difference between heat and temperature?
    Heat is the quantity of energy transferred between objects because of a difference in temperature. Temperature is a measure of the average kinetic energy of the molecules (or other particles) of an object.

Name ________ Date ________ Class ________

Chapter 16

STUDY GUIDE

16.1 CRYSTAL STRUCTURE

1. Complete the table.

Crystal system	Lengths of unit cell axes	Angle between unit cell axes
cubic	equal	all = 90°
tetragonal	2 equal, 1 unequal	all = 90°
orthorhombic	all unequal	all = 90°
monoclinic	all unequal	2 = 90°, 1 ≠ 90°
triclinic	all unequal	all ≠ 90°
rhombohedral	all equal	all ≠ 90°
hexagonal	3 equal, 1 unequal	1 = 90°, 3 = 60°

2. Identify each of the crystal shapes in Figure 1.

Figure 1

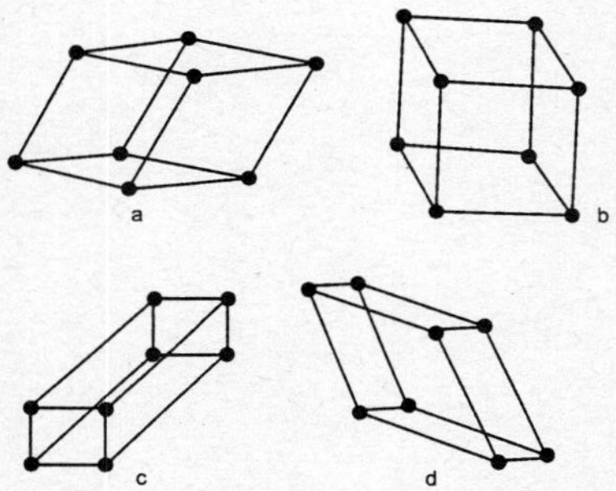

a. rhombohedral
b. simple cubic
c. tetragonal
d. monoclinic

Name ________ Date ________ Class ________

3. Identify each of the unit cells in Figure 2.

Figure 2

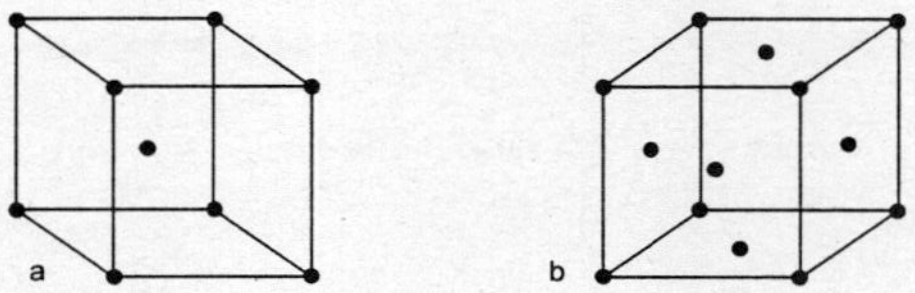

a. body-centered cubic
b. face-centered cubic

Figure 3

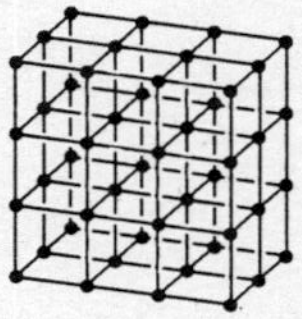

4. a. What is the general name for a structure like the one shown in Figure 3?
   space lattice
   b. If the unit cell is simple cubic, how many unit cells does the structure in Figure 3 contain?
   18

Match each type of bonding with the correct type of crystal.
a. covalent
b. delocalized electrons
c. electrostatic forces
d. van der Waals forces

c 5. ionic
b 6. metallic
a 7. macromolecular/network
d 8. molecular

Name ____________ Date ____________ Class ____________

9. Label each range of melting points on the temperature scale in Figure 4 with the name of the appropriate crystal type.

Figure 4

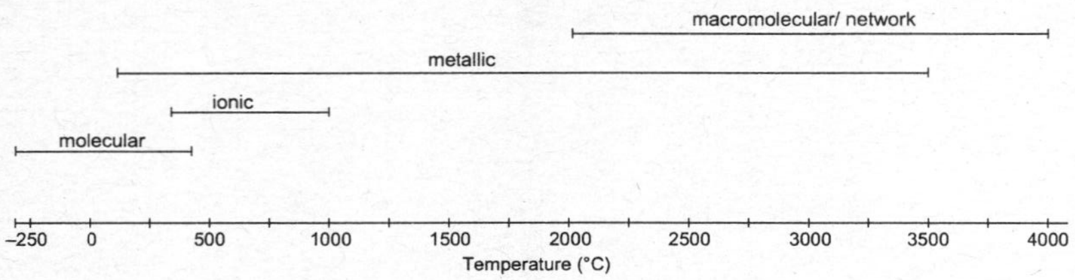
macromolecular/ network
metallic
ionic
molecular
–250 0 500 1000 1500 2000 2500 3000 3500 4000
Temperature (°C)

Write T for true or F for false. If a statement is false, replace the underlined word or phrase with one that will make the statement true, and write your correction on the blank provided.

F; 6 ____ 10. In a crystal of sodium chloride, each Na^+ ion is surrounded by <u>four</u> Cl^- ions.

F; face-centered ____ 11. The unit cell of sodium chloride is <u>body-centered</u> cubic.

F; isomorphous ____ 12. Crystals of different solids with the same structure and shape are said to be <u>polymorphous</u>.

Name ____________ Date ____________ Class ____________

Chapter 16

STUDY GUIDE

16.2 *SPECIAL STRUCTURES*

1. Label the crystal defects shown in Figure 1. Explain how each defect is caused.

Figure 1

edge dislocation: An extra layer of atoms extends part of the way into the crystal.

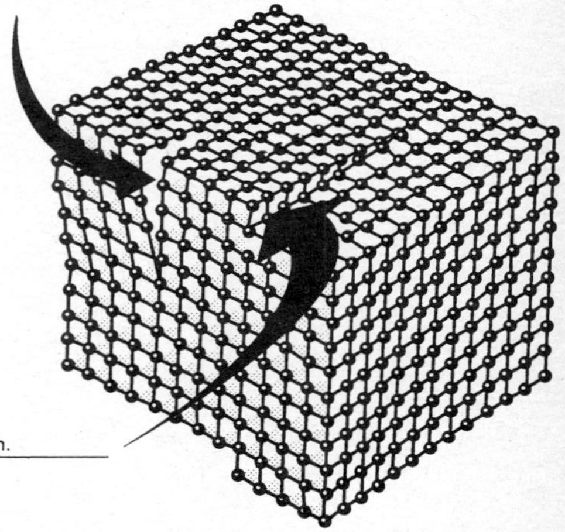

screw dislocation: There is unequal crystal growth.

Write T for true or F for false. If a statement is false, replace the underlined word or phrase with one that will make the statement true, and write your correction on the blank provided.

T ____ 2. Imperfect crystals can be caused by defects to the unit cell, such as <u>misplaced atoms or ions</u>.

T ____ 3. <u>Hydrated ions</u> are chemically bonded to water molecules.

F; hydrated ____ 4. $CuSO_4 \cdot 5H_2O$ and $CaSO_4 \cdot 2H_2O$ are <u>anhydrous</u> compounds.

T ____ 5. It is <u>possible</u> to remove water molecules from hydrated compounds by raising the temperature or lowering the pressure.

F; most ____ 6. Deliquescent materials are the <u>least</u> hygroscopic substances.

T ____ 7. Materials that appear to be solid but that lack crystalline structures are called <u>amorphous</u> materials.

F; (amor) ____ 8. The symbol for amorphous materials is <u>(amph)</u>.

F; higher ____ 9. Glass and molasses have <u>lower</u> viscosities than water and alcohol because they tend to flow less easily.

T ____ 10. Metastable forms of an amorphous substance occur in <u>long-lasting</u> form.

Chapter 17
STUDY GUIDE

17.1 CHANGES OF STATE

Complete the sentence or answer the question.

1. What is formed when molecules of a liquid or a solid substance escape from the surface of the substance? vapor

2. According to kinetic theory, if the temperature of a liquid is lowered, the average velocity of its particles should decrease .

3. When the ordered arrangement of the atoms of a solid substance breaks down, the solid melts .

4. When a liquid freezes, _____ the particles of the substance settle into an ordered arrangement and form a solid.

5. The equation $X(l) \rightarrow X(g)$ represents change from a liquid to a vapor .

6. Write an equation that shows water, H_2O, as it changes reversibly between the liquid and the solid states. $H_2O(l) \rightleftharpoons H_2O(cr)$

7. According to Le Chatelier's principle, what happens if stress is applied to a system at equilibrium? The system tends to readjust so that stress is reduced.

8. Write the name of the device shown in the figure.

 barometer

9. What are the two devices shown in Figure 17.1 used to measure? vapor pressure

10. What is true about the vapor pressures of substances that have strong intermolecular forces? The vapor pressures are low.

11. What is meant by the melting point for a substance? the temperature at which the vapor pressure of the liquid and the vapor pressure of the solid are equal

12. When dry ice is placed in an open container at room temperature, it changes directly from a solid to a vapor. This process is an example of sublimation .

13. The normal boiling point of a substance is the temperature at which its vapor pressure is equal to 101.325 kPa, or standard atmospheric pressure .

Use the letters of the two-dimensional diagrams of (a) crystalline quartz and (b) quartz glass, shown in Figure 2, to answer the following questions.

Figure 2

b 11. Which substance shows the structure of an amorphous material?
a 12. Which substance illustrates long-range order?
a 13. Which substance has a specific melting point?
b 14. Which substance can be classified as a supercooled liquid?
b 15. Which substance is less stable in the form shown?

14. A liquid that boils at a low temperature and evaporates rapidly at room temperature is described as being **volatile**.

15. The temperature above which no amount of pressure will liquefy a gas is the **critical temperature** of the gas.

16. The condensation of substances that are normally gases is called **liquefaction**.

17. Placing a checkmark in the correct column, identify whether each characteristic listed in the table applies to a substance with strong intermolecular forces or to a substance with weak intermolecular forces.

Characteristic	Intermolecular forces	
	Strong	Weak
Volatile		✓
High boiling point	✓	
High evaporation rate		✓
Low vapor pressure at room temperature	✓	
Low critical temperature		✓

18. The specific heat of ice is 2.06 J/g·C°. Calculate how much energy is needed to heat a 52.0-g sample of ice from −20.0°C to 0.0°C.
$q = m \times (\Delta T) C_p$
$q = 52.0 \text{ g} \times 20.0\text{C°} \times \dfrac{2.06 \text{ J}}{\text{g·C°}}$
$q = 2.14 \times 10^3$ J

19. The enthalpy of fusion of ice is 334 J/g. Calculate how much energy is needed to melt a 52.0-g sample of ice.
$q = m \times \Delta H_{\text{fus}}$
$q = 52.0 \text{ g} \times \dfrac{334 \text{ J}}{\text{g}}$
$q = 1.74 \times 10^4$ J

20. The C_p of liquid water is 4.18 J/g·C°. Calculate how much energy is needed to heat a 52.0-g sample of liquid water from 0.0°C to 100.0°C.
$q = m \times (\Delta T) C_p$
$q = 52.0 \text{ g} \times 100.0\text{C°} \times \dfrac{4.18 \text{ J}}{\text{g·C°}}$
$q = 2.17 \times 10^4$ J

21. The enthalpy of vaporization of water is 2260 J/g. Calculate how much energy is needed to boil a 52.0-g sample of liquid water that is already at the boiling point.
$q = m \times \Delta H_{\text{vap}}$
$q = 52.0 \text{ g} \times \dfrac{2260 \text{ J}}{\text{g}}$
$q = 1.18 \times 10^5$ J

22. The C_p of steam is 2.02 J/g·C°. Calculate how much energy is needed to heat 52.0-g of steam from 100.0°C to 120.0°C.
$q = m \times (\Delta T) C_p$
$q = 52.0 \text{ g} \times 20.0\text{C°} \times \dfrac{2.02 \text{ J}}{\text{g·C°}}$
$q = 2.10 \times 10^3$ J

23. Use your calculations from questions 18 through 22 to calculate how much energy is necessary to convert a 52.0-g sample of ice at −20.0°C to steam at 120.0°C.

The total heat that must be absorbed is the sum of the five responses to questions 18-22.
2140 + 17 400 J + 21 700 J + 118 000 J + 2100 J = (rounded) 161 000 J
= 1.61 × 10⁵ J

Chapter 17
STUDY GUIDE

17.2 SPECIAL PROPERTIES

Write T for true and F for false. If a statement is false, replace the underlined word or phrase with one that will make the statement true, and write your correction on the blank provided.

F; expands   1. Water contracts when it freezes.

T   2. Compared with most molecules, water molecules have low masses.

F; liquid   3. Water is a gas at room temperature and standard atmospheric pressure.

T   4. Carbon dioxide and nitrogen are gases at room temperature and standard atmospheric pressure.

T   5. The structure of the molecules in a substance affects the interatomic and intermolecular forces that hold the substance together.

T   6. Molecules that contain hydrogen often have boiling points and melting points that are higher than would be expected in the absence of such hydrogen.

T   7. A molecule that is made up of a highly electronegative atom and a hydrogen atom is highly polar.

F; proton   8. If an actual H+ ion existed, it would consist of a bare neutron.

F; covalently   9. In compounds, hydrogen is always ionically bonded.

F; negative   10. In a substance that is made up of identical polar molecules that contain a hydrogen atom, the hydrogen atom of each molecule in the substance is attracted to the positive portion of the other molecules.

T   11. A hydrogen bond results in a fairly strong hydrogen atom attachment between two molecules.

T   12. The attractive force between hydrogen-bonded molecules is much greater than the attractive force between other dipoles with the same electronegativity difference.

T   13. Hydrogen bonding is one type of dipole attraction.

T   14. Liquid water at 1°C is less dense than liquid water at 3°C.

F; less   15. Because ice is more dense than liquid water, ice floats at the surfaces of lakes and streams.

Match each term with the correct description.

a. hydrogen bonding
b. capillary rise
c. surface tension

c   16. apparent elasticity at the surface of a liquid

b   17. change in elevation of a liquid in a tube with a small diameter

a   18. an attractive force that exists between molecules that contain hydrogen and a highly electronegative atom

19. A water molecule is a dipole. Look at the diagram of the water molecule. Place + and − signs in the blanks to label the areas of the molecule that are partially positive and partially negative.

Chapter 18
STUDY GUIDE

18.1 VARIABLE CONDITIONS

Complete the sentence or answer the question.

1. According to the kinetic theory, a gas is made of very small particles that are in constant random motion.

2. What are the three factors on which the pressure exerted by a gas depends? the number of molecules, the volume they are in, the average kinetic energy of the molecules

3. a. If the number of molecules in a constant volume increases, the pressure __increases__

   b. If the number of molecules and the volume remain constant, but the kinetic energy of the molecules increases, the pressure __increases__

   c. The kinetic energy depends on the __temperature__.

4. Study the graph and answer the questions.

   a. What gas law is illustrated in the graph? Boyle's law

   b. Write a sentence describing the relationship represented by the graph. The volume of a gas varies inversely with its pressure.

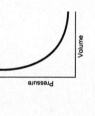

Write T for true or F for false. If a statement is false, replace the underlined word or phrase with one that will make the statement true, and write your correction on the blank provided.

__F; partial__ 5. When a gas is one in a mixture of gases, the pressure exerted by the individual gas is called its __ideal__ pressure.

__F; Kelvin__ 6. In calculations involving gases, the temperature given in Celsius must be converted to __Fahrenheit__.

__T__ 7. At a constant pressure, the volume of a quantity of gas varies __directly__ with the Kelvin temperature.

__F; insoluble__ 8. Gases collected by water displacement must be __soluble__ in water.

__F; Dalton's__ 9. __Charles's__ law is represented by $P_{total} = P_1 + P_2 + \ldots P_n$.

Use Tables 18.1 and 18.2 in the text to answer questions 10 and 11.

10. What is the partial pressure of oxygen in the air at standard conditions? 0.21 × 101.3 kPa = 21 kPa

11. a. What is the vapor pressure of water at 32°C? 4.8 kPa

    b. At what temperature is the vapor pressure of water 2.2 kPa? 19°C

12. Complete the table by using Boyle's law. In the column labeled *Change in volume*, indicate whether the volume increases or decreases. In the next column, write the two possible ratios for pressure and circle the appropriate ratio needed for calculations. Calculate the final volume.

Initial volume	Change in pressure	Change in volume	Pressure ratios	Final volume
26 cm³	55.8 kPa to 110.1 kPa	decrease	55.8 kPa / (110.1 kPa), 110.1 kPa / 55.8 kPa	13 cm³
131 dm³	225 kPa to 650.5 kPa	decrease	225 kPa / (650.5 kPa), 650.5 kPa / 225 kPa	45.3 dm³
88 dm³	36.8 kPa to 22.4 kPa	increase	(36.8 kPa) / 22.4 kPa, 22.4 kPa / 36.8 kPa	140 dm³
925.0 cm³	151.2 kPa to 119.7 kPa	increase	(151.2 kPa) / 119.7 kPa, 119.7 kPa / 151.2 kPa	1168 cm³
0.621 m³	49.5 kPa to 76.2 kPa	decrease	49.5 kPa / (76.2 kPa), 76.2 kPa / 49.5 kPa	0.403 m³

13. A 400-cm³ volume of gas is collected at 26°C. What volume would this gas occupy at standard conditions? Assume a constant pressure.

$$V_2 = 400 \text{ cm}^3 \left| \frac{273 \text{ K}}{299 \text{ K}} \right| = 365 \text{ cm}^3$$

Match each equation below with the correct descriptions.

a. $P_{gas} = P_{total} - P_{water}$ b. $PV = k$ c. $P_{total} = P_1 + P_2 + \ldots P_n$ d. $V = kT$ e. $K = °C + 273.15$

__e__ 14. relationship between the Kelvin and Celsius temperature scales

__b__ 15. Boyle's law

__d__ 16. Charles's law

__c__ 17. Dalton's law

__a__ 18. used to find pressure of a gas collected over water

Chapter 18 STUDY GUIDE

18.2 ADDITIONAL CONSIDERATIONS OF GASES

Answer the questions in the spaces provided.

1. Why is it usually necessary to correct laboratory volumes of gas for both temperature and pressure?
   Laboratory experiments are almost always made at temperatures and pressures other than standard.

2. How are the corrections made?
   by multiplying the original volume by two ratios, one for temperature and the other for pressure

3. Why can the corrections for temperature and pressure be made in either order in one equation?
   The initial volume is multiplied by both ratios, and multiplication is commutative.

4. Why would people across the room from a newly opened bottle of perfume soon be able to smell the perfume?
   As a result of diffusion, perfume vapor molecules would move throughout the room.

5. Based on the diagrams in Figure 1, which gas diffuses faster, ammonia (top row) or carbon dioxide (bottom row)? ammonia

Figure 1

6. Calculate the relative rate of diffusion of the two gases.

   $\dfrac{v_{NH_3}}{v_{CO_2}} = \sqrt{\dfrac{m_{CO_2}}{m_{NH_3}}} = \sqrt{\dfrac{44}{17}} = 1.6$

   Ammonia will diffuse at a rate 1.6 times faster than carbon dioxide.

7. a. Which gas diffuses faster, helium or neon? helium

   b. How much faster?

   $\dfrac{v_{He}}{v_{Ne}} = \sqrt{\dfrac{m_{Ne}}{m_{He}}} = \sqrt{\dfrac{20.2}{4.0}} = 2.2$

   2.2 times faster

8. Using the illustrations for ammonia and carbon dioxide as guides, fill in the blank diagrams in Figure 2 comparing the rates of diffusion of helium and neon. Label your diagrams.

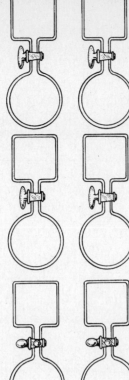

Figure 2

Students' diagrams should show molecules of helium diffusing at a faster rate than molecules of neon (more helium molecules diffused in the second group of flasks). By the third group of flasks, the distribution of the gas particles should be about equal. Be sure students have correctly labeled the flasks of helium and neon.

Chapter 19
STUDY GUIDE

19.1 AVOGADRO'S PRINCIPLE

Complete the sentence or answer the question.

1. What is Avogadro's principle?
   At equal temperatures and equal pressures, equal volumes of gases contain the same number of molecules.

2. Based on Avogadro's principle, if the volumes of two gases under similar conditions are equal, how are their number of moles related?
   The number of moles also would be equal.

3. What is molar volume?
   the volume occupied by 1 mole of any gas under standard conditions, 22.4 dm^3

4. State what each of the five symbols in the ideal gas equation stands for.
   P—pressure; V—volume; n—number of moles; R—constant used in calculations involving the ideal gas equation; T—temperature

5. What temperature scale is used to solve problems involving the ideal gas equation?
   Kelvin

6. Using the ideal gas equation, complete the table by solving for the unknown variable. Show your work below.

P (kPa)	V (dm^3)	n	T (K)
55.6	85.0	6.50	87.5
37.1	210.0	3.20	293
113	8.5	0.37	311
95.1	133	5.50	277

$T = PV/nR = \dfrac{55.6 \text{ kPa}}{6.50 \text{ mol}} \cdot \dfrac{85.0 \text{ dm}^3}{1} \cdot \dfrac{\text{mol} \cdot \text{K}}{8.31 \text{ dm}^3 \cdot \text{kPa}} = 87.5$ K

$P = nRT/V = \dfrac{3.20 \text{ mol}}{210.0 \text{ dm}^3} \cdot \dfrac{8.31 \text{ dm}^3 \cdot \text{kPa}}{\text{mol} \cdot \text{K}} \cdot \dfrac{293 \text{ K}}{1} = 37.1$ kPa

$n = PV/RT = \dfrac{113 \text{ kPa}}{8.31 \text{ dm}^3 \cdot \text{kPa}} \cdot \dfrac{8.5 \text{ dm}^3}{311 \text{ K}} \cdot \dfrac{\text{mol} \cdot \text{K}}{1} = 0.37$ mol

$V = nRT/P = \dfrac{5.50 \text{ mol}}{95.1 \text{ kPa}} \cdot \dfrac{8.31 \text{ dm}^3 \cdot \text{kPa}}{\text{mol} \cdot \text{K}} \cdot \dfrac{277 \text{ K}}{1} = 133$ dm^3

7. In the modified form of the ideal gas equation $PV = mRT/M$, what do the symbols m and M stand for?
   m, mass; M, molecular mass

8. How can this modified equation also be used to solve for the density of a gas?
   Because $D = m/V$, then $D = m/V = PM/RT$.

9. Complete the table by solving for M. Show your work below.

P (kPa)	V (dm^3)	m (g)	T (K)	M (g/mol)
98.6	6.27	0.832	300.0	3.36
112.5	10.1	2.36	298	5.14
88.7	23.5	0.55	292.5	0.64
91.2	2.75	0.124	312	1.28

$M = \dfrac{0.832 \text{ g}}{98.6 \text{ kPa}} \cdot \dfrac{8.31 \text{ dm}^3 \cdot \text{kPa}}{\text{mol} \cdot \text{K}} \cdot \dfrac{300.0 \text{ K}}{6.27 \text{ dm}^3} = 3.36$ g/mol

$M = \dfrac{2.36 \text{ g}}{112.5 \text{ kPa}} \cdot \dfrac{8.31 \text{ dm}^3 \cdot \text{kPa}}{\text{mol} \cdot \text{K}} \cdot \dfrac{298 \text{ K}}{10.1 \text{ dm}^3} = 5.14$ g/mol

$M = \dfrac{0.55 \text{ g}}{88.7 \text{ kPa}} \cdot \dfrac{8.31 \text{ dm}^3 \cdot \text{kPa}}{\text{mol} \cdot \text{K}} \cdot \dfrac{292.5 \text{ K}}{23.5 \text{ dm}^3} = 0.64$ g/mol

$M = \dfrac{0.124 \text{ g}}{91.2 \text{ kPa}} \cdot \dfrac{8.31 \text{ dm}^3 \cdot \text{kPa}}{\text{mol} \cdot \text{K}} \cdot \dfrac{312 \text{ K}}{2.75 \text{ dm}^3} = 1.28$ g/mol

Write T for true or F for false. If a statement is false, replace the underlined word or phrase with one that will make the statement true, and write your correction on the blank provided.

__T__ 10. At a given <u>temperature</u>, the average kinetic energy of all gas molecules is the same.

__F; mole__ 11. One <u>gram</u> of oxygen will occupy 22.4 dm^3 at STP.

__F; Kelvin__ 12. The <u>Celsius</u> temperature scale is used for problems involving the ideal gas equation.

__T__ 13. <u>Density</u> is equal to mass divided by volume.

__T__ 14. The ideal gas equation combines <u>Boyle's</u> law and Charles's law.

Chapter 19
STUDY GUIDE

19.2 GAS STOICHIOMETRY

Complete the sentence or answer the question.

1. What are the four steps you should keep in mind when solving mass-gas volume problems?

   1. Write a balanced equation. 2. Find the number of moles of the given substance. 3. Find the ratio of the moles of the given substance to the moles of required substance. 4. Express moles of gas in terms of volume of gas.

2. What are the four steps you should keep in mind when solving gas volume-mass problems?

   1. Write a balanced equation. 2. Change volume of gas to moles of gas. 3. Determine the ratio of moles of the given substance to moles of required substance. 4. Express moles of given substance as grams of required substance.

Solve each of the following problems.

3. What volume of carbon dioxide, CO_2, at STP can be produced when 8.0 grams of oxygen, O_2, react with an excess of ethane, C_2H_6? Use the steps you listed in questions 1 and 2.

   Step 1: $2C_2H_6 + 7O_2 \rightarrow 4CO_2 + 6H_2O$

   Step 2: $\dfrac{8.0 \text{ g } O_2}{} \Big| \dfrac{1 \text{ mol } O_2}{32 \text{ g } O_2} \ldots$

   Step 3: 7 moles $O_2 \rightarrow$ 4 moles CO_2

   Step 4: $\dfrac{8.0 \text{ g } O_2}{} \Big| \dfrac{1 \text{ mol } O_2}{32 \text{ g } O_2} \Big| \dfrac{4 \text{ mol } CO_2}{7 \text{ mol } O_2} \Big| \dfrac{22.4 \text{ dm}^3}{1 \text{ mol}} = 3.2 \text{ dm}^3 \, CO_2$

4. a. How many grams of carbon dioxide, CO_2, are formed if 96.0 g of oxygen, O_2, react with 12.2 dm³ of ethylene, C_2H_4, to form carbon dioxide and water? Assume STP. Which reactant is the limiting one?

   $C_2H_4 + 3O_2 \rightarrow 2CO_2 + 2H_2O$

   $\dfrac{96.0 \text{ g } O_2}{} \Big| \dfrac{1 \text{ mol } O_2}{32.0 \text{ g } O_2} = 3 \text{ mol } O_2$

   $\dfrac{12.2 \text{ dm}^3 \, C_2H_4}{} \Big| \dfrac{1 \text{ mol}}{22.4 \text{ dm}^3} = 0.54 \text{ mol } C_2H_4$ (limiting reactant)

   $\dfrac{0.54 \text{ mol } C_2H_4}{} \Big| \dfrac{2 \text{ mol } CO_2}{1 \text{ mol } C_2H_4} \Big| \dfrac{44.0 \text{ g } CO_2}{1 \text{ mol } CO_2} = 48 \text{ g } CO_2$

b. How many grams of CO_2 are produced when 20.0 g of O_2 react with 31 dm³ of ethylene? Assume STP. Which reactant is the limiting one?

$C_2H_4 + 3O_2 \rightarrow 2CO_2 + 2H_2O$

$\dfrac{20.0 \text{ g } O_2}{} \Big| \dfrac{1 \text{ mol } O_2}{32 \text{ g } O_2} = 0.63 \text{ mol } O_2$ (limiting reactant)

$\dfrac{31 \text{ dm}^3 \, C_2H_4}{} \Big| \dfrac{1 \text{ mol}}{22.4 \text{ dm}^3} = 1.4 \text{ mol } C_2H_4$

$\dfrac{0.63 \text{ mol } O_2}{} \Big| \dfrac{2 \text{ mol } CO_2}{3 \text{ mol } O_2} \Big| \dfrac{44.0 \text{ g } CO_2}{1 \text{ mol } CO_2} = 18 \text{ g } CO_2$

Write T for true or F for false. If a statement is false, replace the underlined word or phrase with one that will make the statement true, and write your correction on the blank provided.

__T__ 5. There is more than one way to find out which reactant is the <u>limiting</u> reactant in a chemical reaction.

__T__ 6. <u>Volume-volume</u> relationships can be used when all the reactants and products are gases.

__T__ 7. It is usually awkward to measure the <u>mass</u> of a gas.

__F: mole__ 8. A balanced equation tells you the <u>volume</u> ratios of the reactants and products.

__F: limiting__ 9. In a chemical reaction, the amount of products is determined by the <u>excess</u> reactant.

Solve the following problems.

10. In a reaction between hydrogen and bromine to form hydrogen bromide gas, HBr, how many dm³ of H_2 will be needed to completely use up 65.0 dm³ of Br_2? (Assume constant pressure and temperature.)

    $H_2 + Br_2 \rightarrow 2HBr, 1 \text{ mol } H_2 \rightarrow 1 \text{ mol } Br_2$

    $\dfrac{65.0 \text{ dm}^3 \, Br_2}{} \Big| \dfrac{1 \text{ mol } H_2}{1 \text{ mol } Br_2} = 65.0 \text{ dm}^3 \, H_2$

11. How many moles and how many grams of HBr will be produced when the hydrogen and bromine react?

    $\dfrac{65.0 \text{ dm}^3 \, Br_2}{} \Big| \dfrac{1 \text{ mol}}{22.4 \text{ dm}^3} \Big| \dfrac{2 \text{ mol HBr}}{1 \text{ mol } Br_2} = 5.80 \text{ mol HBr}$

    $\dfrac{5.80 \text{ mol HBr}}{} \Big| \dfrac{80.9 \text{ g HBr}}{1 \text{ mol HBr}} = 469 \text{ g HBr}$

12. If only 22.4 dm³ of H_2 reacted with 65 dm³ of Br_2, how many grams of HBr would be produced?

    $\dfrac{22.4 \text{ dm}^3 \, H_2}{} \Big| \dfrac{1 \text{ mol}}{22.4 \text{ dm}^3} \Big| \dfrac{2 \text{ mol HBr}}{1 \text{ mol } H_2} \Big| \dfrac{80.9 \text{ g HBr}}{1 \text{ mol HBr}} = 162 \text{ g HBr}$

Chapter 20
STUDY GUIDE

20.1 SOLUTIONS

Write T for true or F for false. If a statement is false, replace the underlined word or phrase with one that will make the statement true, and write your correction on the blank provided.

F; solid   1. Most solutions consist of a <u>gas</u> dissolved in a liquid.

F; molarity 2. <u>Molality</u> is the most common concentration unit in chemistry.

T   3. <u>Water</u> is the most common solvent.

F; alloy   4. An <u>amalgam</u> is a solid metal-metal solution.

T   5. The solubility of a substance can be changed by altering the <u>temperature</u>.

F; nine   6. In terms of physical states of matter, there are <u>three</u> possible combinations of solvent-solute pairs.

F; negative 7. Gases have <u>positive</u> enthalpies of solution.

T   8. The differing <u>solubilities</u> of substances can be used to separate them from mixtures.

T   9. Solvation <u>does not</u> occur in the case of a polar solvent and a nonpolar solute.

F; endothermic 10. The solution process is usually <u>exothermic</u>.

Answer the questions below.

11. Why does a solution of potassium chloride not exhibit the characteristic behavior of potassium chloride?

    There is really no potassium chloride in solution. Dissociated ions act as if they were present alone.

12. How is supersaturation possible?

    If a solvent is heated, additional solute will dissolve. When the solution is cooled, the solute will remain in the solution so long as there is no angular surface upon which to start crystallization. A container that has a smooth interior and contains a dust-free solute has no such surfaces.

13. What are three things that you could do to speed up the rate of solution?

    Answers may include: break the solute into small pieces, stir the solution, and heat the solution.

14. What effect does pressure have on solutions in which the solute is a gas?

    The higher the pressure is, the greater is the amount of gas that will dissolve in a given amount of solvent.

15. What is the difference between molarity and molality?

    Molarity expresses concentration as moles of solute per dm^3 of solution. Molality expresses concentration as moles of solute per kilogram of solvent.

16. In terms of polarity, which solvent-solute combinations are most likely to form solutions?

    Polar solutes most likely form solutions with polar solvents. Nonpolar solutes most likely form solutions with nonpolar solvents.

17. Complete the table shown as a review of the solubility rules for water solutions. For each substance listed, place a check mark in the correct column indicating whether you would predict it to be soluble or insoluble in water. (Use Appendix Table A-7 for reference.)

Substance	Soluble	Insoluble
$Al_2(SO_4)_3$	x	
$PbBr_2$		x
$ZnCO_3$		x
$CaBr_2$	x	
$(NH_4)_3PO_4$	x	
$Mn(NO_3)_2$	x	
$SrSO_4$		x
MgF_2		x
$Co(CH_3COO)_3$	x	

18. Match each substance a-c with the reagent(s) that can be used to precipitate the positive ion. Write your answer(s) on the lines provided.

    Reagents: K_3PO_4, LiI, $Al_2(SO_4)_3$

    a. $SrBr_2$ K_3PO_4, $Al_2(SO_4)_3$
    b. $MgCl_2$ K_3PO_4
    c. $AgNO_3$ K_3PO_4, LiI, $Al_2(SO_4)_3$

Solve the following concentration problems.

19. If 10.0 g of sodium hydroxide is dissolved in 3.0 dm^3 of water, what is the molarity of the NaOH solution formed?

    $$\frac{10.0 \text{ g NaOH}}{3.0 \text{ dm}^3} \left| \frac{1 \text{ mol NaOH}}{40.0 \text{ g NaOH}} \right. = 0.083 M$$

20. What is the molality of a solution formed by dissolving 60.0 g of HNO_3 in 2.50×10^3 g of water?

    $$\frac{60.0 \text{ g HNO}_3}{2.50 \times 10^3 \text{ g H}_2\text{O}} \left| \frac{1 \text{ mol HNO}_3}{63.0 \text{ g HNO}_3} \right| \frac{1000 \text{ g H}_2\text{O}}{1 \text{ kg H}_2\text{O}} = 0.381 m$$

21. If 150 g of n-butylacetate, $C_6H_{12}O_2$, is dissolved in 190 g of ethanol, C_2H_6O, what is the mole fraction of each component of the solution?

    moles of n-butyl acetate $= \dfrac{150 \text{ g}}{} \left| \dfrac{1 \text{ mol}}{116 \text{ g}} \right. = 1.3 \text{ mol}$ mole fraction of n-butyl acetate $= \dfrac{1.29 \text{ mol}}{5.4 \text{ mol}} = 0.24$

    moles of ethanol $= \dfrac{190 \text{ g}}{} \left| \dfrac{1 \text{ mol}}{46 \text{ g}} \right. = 4.1 \text{ mol}$ mole fraction of ethanol $= \dfrac{4.1 \text{ mol}}{5.4 \text{ mol}} = 0.76$

Chapter 20
STUDY GUIDE
20.2 COLLOIDS

Match each type of colloid a through g with the correct example from the list that follows. Write the correct letter on the line provided. Some answers may be used more than once.

a. liquid emulsion   d. aerosols   g. liquid sols
b. solid foam   e. solid emulsion
c. solid sols   f. liquid foam

a 1. mayonnaise
f 2. whipped cream
c 3. opals
e 4. cheese
g 5. jelly
d 6. smoke
c 7. pearls
d 8. fog
g 9. paint
b 10. marshmallows

11. Complete the table comparing particles in solutions, suspensions, and colloids. Write *yes* or *no* for each characteristic to indicate whether it describes solutions, suspensions, or colloids.

Characteristic	Solutions	Suspensions	Colloids
less than 1 nm	yes	no	no
greater than 100 nm	no	yes	no
greater than 1 nm, but less than 100 nm	no	no	yes
settle out on standing	no	yes	no
pass unchanged through ordinary filter paper	yes	no	yes
pass unchanged through membranes	yes	no	no
scatter light	no	yes	yes
affect colligative properties	yes	no	no
particles can be seen with an ordinary light microscope	no	yes	no
heterogeneous	no	yes	yes

Answer each of the following questions on the lines provided.

12. What is colloid chemistry?

    the study of the properties of matter whose particles are between 1 nm and 100 nm in at least one dimension

13. Why are colloids frequently used as the stationary phase in chromatography?

    The stationary phase in chromatography is an adsorbing material. Colloidal particles have a large ratio of surface area to mass, which makes them excellent absorbing materials.

14. Why can semipermeable membranes be used to separate ions and colloidal particles?

    The ions pass through the membranes but the colloidal particles do not because they are too large.

Match each property of colloids with the letter of the correct description. Write your answer on the line provided.

a. random motion of colloidal particles
b. scattering of a beam of light
c. attraction and holding of substances on the surfaces of dispersed particles
d. migration of positive and negative colloidal particles in an electric field

b 15. Tyndall effect
c 16. adsorption
a 17. Brownian motion
d 18. electrophoresis

19. The Figures 1 and 2 illustrate two characteristic properties of colloids. In each case, identify the property and explain why it occurs.

Figure 2

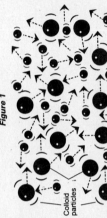

Figure 1

Figure 1: Brownian motion; The colloidal particles are constantly bombarded by the smaller molecules of the medium. The motion is the result of the collision of many molecules with the particles.

Figure 2: Tyndall effect; A beam of light passing through a colloid is scattered by the colloidal particles and becomes visible from the side.

Chapter 21
STUDY GUIDE

21.1 VAPOR PRESSURE CHANGES

For each of the following, write T if the statement is true. If the statement is false, replace the underlined portion with a word or phrase that makes the statement true. Write your answer on the blank provided.

colligative 1. Vapor pressure, freezing point, and boiling point are <u>chemical</u> properties of a solution.

number 2. Colligative properties depend on the <u>kind</u> of particles present in solution.

T 3. If fewer solvent particles can evaporate from a solution, the vapor pressure of the solvent is <u>lowered</u>.

solvent 4. For a solution having a specific concentration, the amount that the vapor pressure of the <u>solvent</u> is lowered depends on the characteristics of the <u>solute</u>.

nonvolatile 5. Glucose, sucrose, and other molecular compounds with high melting points are considered <u>volatile</u> solutes.

T 6. The amount the vapor pressure of a solvent is lowered is equal to the difference between the vapor pressure of the pure solvent and the vapor pressure of the <u>solution</u>.

solvent 7. According to Raoult's law, the vapor pressure of a solution is directly proportional to the mole fraction of <u>solute</u>.

3/4 or 0.75 8. The vapor pressure of a sugar solution in which the mole fraction of sugar is 0.25 will be <u>4</u> times the vapor pressure of pure water.

T 9. In an ideal solution, all possible attractions among particles of solute and solvent are <u>the same</u>.

T 10. In an aqueous sucrose solution at 0°C, the mole fraction of water is 0.85. If the vapor pressure of water at 0°C is 0.6 kPa, its vapor pressure is lowered by 0.09 kPa.

Use the diagram of a fractional distillation apparatus shown here to answer the following questions about a solution consisting of pure liquids X and Y, where the boiling point of liquid X is greater than that of liquid Y.

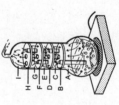

11. Which of the substances in the given solution is more volatile? Why? <u>substance Y; It has a lower boiling point.</u>

---

12. When the vapor at A is in equilibrium with the given solution, is the vapor richer in substance X or substance Y? <u>Y</u>

13. The substances at D and E are in equilibrium. How does the temperature of the solution at D compare with the temperature of the vapor at E? <u>The temperatures are the same.</u>

14. How does the temperature at I compare with the temperature at A? Explain your reasoning. <u>The temperature at I is lower than the temperature at A because I is farther from the heat source.</u>

15. How does the composition of the substances at the top of the apparatus (H and I) differ from the composition of the substances at B and C? <u>The concentration of substance Y is much greater at H and I than at B and C.</u>

Use Table 21.1 in your textbook to answer the following questions about the fractional distillation of petroleum.

16. Which fraction(s) would come off first? Explain your reasoning. <u>gasoline; It has the lowest boiling point range and, therefore, is more volatile than the other fractions.</u>

17. Which fraction(s) would collect at the bottom of the fractionating tower? <u>lubricating oil</u>

18. Would the fractions at the middle levels of the tower be richer in fuel oil or jet fuel? <u>jet fuel</u>

Use the symbol > (greater than) or < (less than) to complete each of the following comparisons between vapor pressures, boiling points, and freezing points of (1) a pure solvent and (2) a solution of a nonvolatile solute in the solvent at the same temperature.

19. vapor pressure: pure solvent <u>></u> solution
20. boiling point: pure solvent <u><</u> solution
21. freezing point: pure solvent <u>></u> solution

Answer each of the following questions.

22. Describe and explain the effect of adding a nonvolatile, nonionizing solute on the boiling point of a pure solvent. <u>A solution boils when the vapor pressure of the liquid equals atmospheric pressure. The addition of a solute lowers the vapor pressure of the solvent in the solution. A higher temperature than is required to put enough solvent particles into the vapor phase to equal atmospheric pressure.</u>

23. Why is the addition of antifreeze to the water in a car radiator just as important in hot climates as in cold climates? <u>A solution of antifreeze and water has a higher boiling point than that of water or antifreeze alone, thus preventing the radiator from boiling over on very hot days.</u>

Chapter 21
STUDY GUIDE

21.2 QUANTITATIVE CHANGES

Answer each of the following questions.

1. What is meant by the molal boiling point constant and molal freezing point constant for a given solvent?

   Molal boiling point constant is the number of C° that the boiling point of 1 kg of the solvent is raised when 1 mole of nonvolatile solute particles is added. Molal freezing point constant is the number of C° that the freezing point of 1 kg of the solvent is lowered when 1 mole of nonvolatile solute particles is added.

2. Describe freezing point depression in terms of the freezing points of a pure solvent and a solution made from that solvent.

   Freezing point depression is equal to the difference in temperature between the freezing point of the pure solvent and the freezing point of the solution.

3. Ideally, a $1m$ solution of zinc chloride ($ZnCl_2$) in water would boil at 101.545°C, whereas a $1m$ solution of glucose ($C_6H_{12}O_6$) would boil at 100.515°C.

   a. What is the boiling point elevation of an ideal solution of zinc chloride in water?

   1.545 C°

   b. Compare the boiling point elevation of a zinc chloride solution with the boiling point elevation of a glucose solution.

   The boiling point elevation of zinc chloride is three times greater than that of glucose.

   c. How does the difference in boiling point elevation provide evidence that zinc chloride dissociates in water?

   Because the boiling point is raised by three times as much, there must be three times as many particles in the zinc chloride solution. Those particles must come from the dissociation of zinc chloride into Zn^{2+} ions and twice as many Cl^- ions.

4. The molal freezing point constant for pure water is 1.853 C°/m. Assuming 100% dissociation, at what temperature would a $1m$ solution of magnesium chloride ($MgCl_2$) in water freeze?

   $3m$ (1.853 C°/m) = 5.559 C°; freezing point = −5.559°C

5. A solution contains 20.0 g of maltose ($C_{12}H_{22}O_{11}$), a nonvolatile, nonionizing solute, dissolved in 324 g H_2O. Fill in the missing parts and complete the following calculations for the boiling and freezing points of the resulting solution.

   a. $\dfrac{20.0 \text{ g } C_{12}H_{22}O_{11}}{324 \text{ g } H_2O} \Big| \dfrac{1 \text{ mol } C_{12}H_{22}O_{11}}{342 \text{ g } C_{12}H_{22}O_{11}} \Big| \dfrac{1000 \text{ g } H_2O}{1 \text{ kg } H_2O} =$ 0.180m

   b. boiling point elevation = ($\dfrac{0.180m}{100.093}$)(0.515 C°/m) = 0.093 C°

      boiling point = 100.093 °C

   c. freezing point depression = (0.180m)(1.853 C°/m) = 0.334 C°

      freezing point = −0.334 °C

In the diagrams shown, a semipermeable membrane covers the opening of the test tube that is inverted in the beaker. The molecules of solvent used can pass through the membrane, but the molecules of solute cannot. Study the diagrams, then answer the questions that follow.

Key: molecules of solvent · molecules of solute

6. Draw arrows on diagram A to indicate the direction and rate of movement of solvent molecules. Explain the reasoning for what you drew.

   Only solvent molecules are present, and they pass at equal rates in both directions.

7. Describe the direction and rate of movement of solvent and solute molecules in diagram B. Draw arrows to represent the movement of solvent molecules.

   More solvent molecules pass from the beaker into the test tube and at a greater rate than solvent molecules pass from the test tube into the beaker. The solute particles do not pass through the membrane.

8. In terms of solvent concentration, describe the net movement of molecules in diagram B.

   Solvent molecules move from higher concentration in the beaker to lower concentration in the test tube.

9. On which side of the membrane in diagram B will an increase in volume take place?

   on the solution side (in the test tube)

10. At equilibrium, on which side of the membrane will osmotic pressure be exerted?

    on the solution side

11. On diagram C, show the solution and the pure solvent at equilibrium. Use arrows to indicate the flow of solvent molecules.

For each of the following, write T if the statement is true. If the statement is false, replace the underlined portion with a word or phrase that makes the statement true. Write your answer on the line provided.

T   12. Ion activity is an indication of the degree to which an ionic compound <u>actually</u> dissociates in a water solution.

attractive; opposite or unlike   13. In reality, ionizing ionic compounds in aqueous solution do not completely dissociate because of the <u>repulsive</u> forces between <u>like</u> charged ions.

more   14. For an ionizing ionic compound dissolved in water, the more dilute the solution, the <u>less</u> effective each ion is in freezing point depression and boiling point elevation.

Chapter 22
STUDY GUIDE

22.1 REACTION RATES

Write T for true or F for false. If the statement is false, replace the underlined word or phrase with one that makes the statement true and write your correction on the blank provided.

thermodynamically stable 1. A substance that does not break down spontaneously at room temperature is said to be kinetically unstable.

T 2. A reaction that eventually reaches equilibrium is said to be reversible.

T 3. A reaction goes to completion if at least one of the reactants is used up.

appears 4. The rate of a reaction from left to right is the rate at which a product disappears.

activation energy 5. Heat, a flame, or a spark may supply the activated complexes needed for a reaction to take place.

T 6. The reactions of kinetically stable substances tend to have high activation energies.

potential 7. The energy required to form an activated complex is converted to kinetic energy.

T 8. Ionic reactions tend to have faster reaction rates than electron transfer reactions.

concentrations 9. Reaction rate varies directly as the product of the masses of the reactants.

T 10. Specific rate constant refers to the rate at which a reaction occurs at a given temperature.

greater than 11. For a given reaction, the total number of molecules having the required activation energy at 125°C would be the same as the total number of molecules having the required activation energy at 100°C.

a catalyst 12. A substance that affects reaction rate without being chemically changed at the end of the reaction is an inhibitor.

Match each situation in questions 13-18 with the letter of the term that best describes it. A letter may be used more than once.

a. homogeneous catalyst   e. inhibitor
b. homogeneous reaction   f. kinetically stable
c. heterogeneous catalyst   g. thermodynamically stable
d. heterogeneous reaction

b 13. $H_2O(g) + CO(g) \rightarrow H_2(g) + CO_2(g)$

f 14. The reaction between $N_2(g)$ and $O_2(g)$ does not take place at room temperature because high activation energies are required.

d 15. $CaO(cr) + CO_2(g) \rightarrow CaCO_3(cr)$

c 16. Manganese dioxide powder is added to increase the decomposition rate of liquid potassium chlorate.

e 17. BHT is added to food as a preservative.

b 18. $C_2H_5OH(aq) + CH_3COOH(aq) \rightleftarrows CH_3COOC_2H_5(aq) + H_2O(l)$

For each of the following reactions, write I if the effect of the given change is an increase in reaction rate, D if the effect is a decrease in reaction rate, or RS if the reaction rate remains the same. Assume that the reaction takes place in a closed vessel of fixed volume.

19. Limestone (calcium carbonate) reacts with hydrochloric acid in an irreversible reaction, to form carbon dioxide and water as described by the following equation:
$CaCO_3(cr) + 2HCl(aq) \rightarrow CaCl_2(aq) + CO_2(g) + H_2O(l)$
What is the effect if

D   a. the temperature is lowered?
RS b. the volume of the reaction vessel is increased?
I   c. limestone chips are used instead of a block of limestone?
RS d. the pressure inside the reaction vessel is increased?
D   e. a more dilute solution of HCl is used?

20. At a given temperature, ethene, C_2H_4, reacts with chlorine in a reversible reaction, to produce 1,2-dichloroethane $(C_2H_4Cl_2)$ as described by the following equation:
$C_2H_4(g) + Cl_2(g) \rightarrow C_2H_4Cl_2(g)$
What is the effect if

I   a. the volume of the reaction vessel is reduced?
D   b. the pressure inside the reaction vessel is reduced?
I   c. a catalyst is added?
D   d. the volume of the reaction vessel is doubled and the pressure inside the vessel is halved?

Use the energy diagram shown to answer questions 21 - 29.

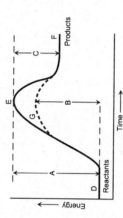

21. Which letter indicates the energy content of the reactants? D
22. Which letter indicates the energy content of the products? F
23. Which letter indicates the change in energy required for the formation of activated complexes in the forward reaction? A
24. Assume that the reaction is reversible. Which line indicates the change in energy required for the formation of activated complexes in the reverse reaction? C
25. Which letter represents the energy of the activated complexes in the uncatalyzed reaction? E
26. Which letter indicates the reaction in the presence of a catalyst? G
27. Which letter indicates the activation energy needed for the forward reaction if a catalyst is used? B

Chapter 22
STUDY GUIDE

22.2 CHEMICAL EQUILIBRIUM

For items 1-8, underline the term inside the parentheses that makes each statement true.

1. At equilibrium, the rate of the forward reaction is (<u>equal to</u>, greater than) the rate of the reverse reaction.

2. The equilibrium constant for a given reaction at a given temperature is the (product, <u>quotient</u>) of the specific rate constant for the forward reaction and the specific rate constant for the reverse reaction.

3. The exponents used in the expression for the equilibrium constant are the (subscripts, <u>coefficients</u>) of the reactants and products.

4. If a reaction tends to go toward completion, the rate of the forward reaction is (equal to, <u>greater than</u>) the rate of the reverse reaction before equilibrium is reached.

5. If $K_{eq} = 1.2 \times 10^{-5}$, the concentration of the reactants is (<u>greater than</u>, less than) the concentration of the products at equilibrium.

6. At temperature T_1, K_{eq} for a certain reaction is 0.239. At temperature T_2, K_{eq} for the same reaction is 4.7. By changing the temperature from T_1 to T_2, the equilibrium will shift in favor of the (reactants, <u>products</u>).

7. If the reaction $H_2(g) + Cl_2(g) \rightleftarrows 2HCl(g) + heat$ is at equilibrium, a decrease in (volume, <u>temperature</u>) will produce a shift in equilibrium toward the right.

8. An increase in pressure on the system $2CO_2(g) \rightleftarrows 2CO(g) + O_2(g)$ at equilibrium results in an equilibrium shift toward the (<u>left</u>, right).

Answer or complete each of the following items.

9. What factors can affect the equilibrium of a reaction?
   temperature, pressure, concentration of reactants, concentration of products

10. A reversible one-step reaction occurs between carbon monoxide gas, CO, and hydrogen gas, H_2, to produce methane gas, CH_4, and gaseous water. Using this information, fill in the diagram according to the following guidelines:
    a. Within the ovals, write the balanced equation for the reaction.
    b. Label the forward and reverse reactions on the lines provided.
    c. In the rectangles, write the words *reactants* or *products*, as appropriate, to represent both the forward and reverse reactions.

```
                    forward
                   ─────────── reaction
   ┌──────────┐       ↑        ┌──────────┐
   │ reactants│               │ products │
   └──────────┘   ⎛ CH₄(g) + H₂O(g) ⎞
        ⇅        ⎝                 ⎠
   ⎛ CO(g) + 3H₂(g) ⎞
   ⎝                ⎠       ↓
   ┌──────────┐              ┌──────────┐
   │ products │              │ reactants│
   └──────────┘              └──────────┘
                   ─────────── reaction
                    reverse
```

28. How are the activation energies of the forward and reverse reactions affected by the addition of a catalyst?
    They are lowered.

29. How does the potential energy of the activated complexes of the forward reaction compare with the potential energy of the activated complexes of the reverse reaction?
    They are the same.

30. For the reaction $A + B \rightarrow C$, the following data were obtained:

Trial	[A]	[B]	Rate
1	0.10M	0.20M	0.00003 (mol/dm³)/s
2	0.10M	0.40M	0.00006 (mol/dm³)/s
3	0.20M	0.40M	0.00048 (mol/dm³)/s

a. How is the rate of reaction affected by a change in concentration of reactant A? Reactant B?
   If [A] doubles, the rate of reaction increases by a factor of 8. If [B] doubles, the rate of reaction doubles.

b. If the rate expression for this reaction is: rate = $k[A]^3[B]$, find k.

   $0.00003 = k(0.10)^3(0.20)$

   $k = \dfrac{0.00003}{(0.10)^3(0.20)} = 0.15$

Chapter 23
STUDY GUIDE

23.1 ACIDS AND BASES

Complete the sentence or answer the question.

1. Define or explain the following theories about acids and bases.

   a. Arrhenius Theory _The Arrhenius Theory states that acids are substances that ionize in water solution to produce hydrogen ions, H^+, or free protons. Bases are substances that ionize in water solution to produce hydroxide ions, OH^-._

   b. Brönsted-Lowry Theory _The Brönsted-Lowry Theory states that in a chemical reaction, any substance that donates a proton is an acid, and any substance that accepts a proton is a base._

2. A(n) _electrolyte_ is an acid, base, or salt, which, when dissolved in water, conducts an electric current.

3. The polyatomic ion, H_3O^+, which is formed when a hydrogen ion (H^+) combines with a water molecule, is called the _hydronium ion_.

4. The _conjugate base_ of an acid is the particle that remains after a proton has been released by the acid.

5. The _conjugate acid_ of a base is formed when the base acquires a proton from the acid.

6. Complete the following equation that, *in general*, describes any acid-base reaction:

   acid + base → _conjugate base_ + _conjugate acid_

7. Acids that contain only two elements are called _binary acids_.

8. Acids that contain three elements are called _ternary acids_.

9. Substances that can react as either an acid or a base are said to be _amphoteric_.

10. The most common example of the kind of compound described in exercise 9 is _water_.

Match each term with the correct description.

a. acidic anhydride   d. strong acid
b. basic anhydride   e. strong base
c. weak acid   f. weak base

_e_ 11. a base that is completely dissociated in solution into positive ions and negative ions
_a_ 12. any oxide that will produce an acid when dissolved in water
_d_ 13. an acid that is considered to ionize completely in solution into positive ions and negative ions
_b_ 14. any oxide that will produce a base when dissolved in water
_c_ 15. slightly ionized acid in solution
_f_ 16. a base that is only partially ionized in solution

---

11. Write the expression for the equilibrium constant for the reaction in question 10.

    $K_{eq} = \dfrac{[CH_4][H_2O]}{[CO][H_2]^3}$

12. If temperature and pressure of the reaction in question 10 are kept constant, but the concentration of each of the substances is halved,

    a. how does the rate of the forward reaction change?
    It decreases by a factor of 16.

    b. how does the rate of the reverse reaction change?
    It decreases by a factor of 4.

    c. what would be the net change in the relative amounts of reactants and products?
    There would be an increase in amount of reactants.

13. In the reaction in question 10, if the volume of the reaction vessel and temperature are kept constant and the pressure on the system is increased,

    a. the concentrations of which substances would be affected? _all_

    b. in which direction will the equilibrium shift? _toward the right_

    c. which substance(s) will show an increase in concentration when equilibrium is reestablished?
    _CH_4, H_2O_

14. Consider the equilibrium equation for the reaction: $4HCl(g) + O_2(g) + heat \rightleftarrows 2Cl_2(g) + 2H_2O(g)$

    a. If the temperature is increased, the reaction will favor the formation of _the forward reaction_

    b. Which reaction requires an input of energy? _the forward reaction_

15. In the Haber process, which involves the reaction $N_2(g) + 3H_2(g) \rightleftarrows 2NH_3 + energy$,

    a. why is NH_3 removed as it is formed?
    The removal of product shifts equilibrium toward the formation of more product.

    b. why is the use of a catalyst considered one of the optimum conditions for this process?
    It lowers the required activation energy, thus speeding up the process.

    c. what is the effect on the relative amounts of product and reactant if the catalyst is removed?
    none

16. Given $K_{eq} = \dfrac{[NO]^6[H_2O]^6}{[NH_3]^4[O_2]^5}$

    a. Write the chemical equation for the reversible reaction having the given K_{eq}.

    $4NH_3 + 5O_2 \rightleftarrows 4NO + 6H_2O$

    b. At a certain temperature the concentrations of NO and NH_3 are equal, and the concentration of H_2O and O_2 are 2.0M and 3.0M respectively. What is the value of K_{eq} at this temperature? When the concentrations of NO and NH_3 are equal, the value of $\dfrac{[NO]}{[NH_3]^4}$ equals one.

    Therefore, $K_{eq} = \dfrac{[2.0]^6}{[3.0]^5}|_1 = \dfrac{64}{243} = 0.26$

Complete the following exercises.

17. Name the following binary acids.
    a. HF _hydrofluoric acid_
    b. HCl _hydrochloric acid_
    c. HI _hydroiodic acid_
    d. H$_2$S _hydrosulfuric acid_
    e. HN$_3$ _hydroazoic acid_
    f. HBr _hydrobromic acid_

18. Name the following ternary acids and bases.
    a. H$_3$AsO$_4$ _arsenic acid_
    b. H$_3$AsO$_3$ _arsenous acid_
    c. NaOH _sodium hydroxide_
    d. H$_3$BO$_3$ _boric acid_
    e. H$_3$PO$_4$ _phosphoric acid_
    f. H$_3$PO$_3$ _phosphorous acid_
    g. HNO$_3$ _nitric acid_

19. Name the following organic acids and bases.
    a. CH$_3$NH$_2$ _methanamine_
    b. CH$_3$COOH _acetic acid (or ethanoic acid)_
    c. CH$_3$CH$_2$COOH _propanoic acid_

20. Write the formula for the conjugate base of each of the acids listed.
    a. H$_2$O _OH$^-$_
    b. HNO$_3$ _NO$_3^-$_
    c. HF _F$^-$_
    d. HC$_2$H$_3$O$_2$ _C$_2$H$_3$O$_2^-$_

21. Write the formula for the conjugate acid of each of the following bases listed.
    a. NH$_3$ _NH$_4^+$_
    b. HSO$_4^-$ _H$_2$SO$_4$_
    c. HS$^-$ _H$_2$S_
    d. C$_2$H$_5$NH _C$_2$H$_5$NH$_2$_

22. Classify each of the following as a strong acid, strong base, weak acid, or weak base.
    a. NaOH _strong base_
    b. HCl _strong acid_
    c. NH$_4^+$ _weak acid_
    d. NH$_3$ _weak base_
    e. Cl$^-$ _weak base_
    f. HI _strong acid_

23. In the following reaction, which species behave as Brønsted acids? As Brønsted bases?
    H$_2$SO$_4$(aq) + H$_2$O $\rightleftharpoons$ HSO$_4^-$(aq) + H$_3$O$^+$(aq)
    Brønsted acids: H$_2$SO$_4$, H$_3$O$^+$; Brønsted bases: H$_2$O, HSO$_4^-$

24. Write the formulas for the anhydrides of the following.
    a. H$_2$CO$_3$ _CO$_2$_
    b. H$_2$SO$_3$ _SO$_2$_
    c. H$_2$SO$_4$ _SO$_3$_
    d. NaOH _Na$_2$O_
    e. Ca(OH)$_2$ _CaO_
    f. HNO$_3$ _N$_2$O$_5$_

Chapter 23
STUDY GUIDE

23.2 SALTS AND SOLUTIONS

Complete the sentence or answer the question.

1. Define or explain the following terms.
   a. salt a crystalline compound composed of the negative ion of an acid and the positive ion of a base

   b. neutralization reaction the reaction of an acid and a base

   c. polyprotic acid an acid containing more than one ionizable hydrogen atom

2. Ions present in a solution but not involved in the reaction are called spectator ions

3. State the general rule that must be observed when writing net ionic equations.
   Substances occurring in a reaction in molecular form are written as molecules. Substances occurring as ions are written as ions.

4. Convert the following balanced equations first to ionic form and then to net ionic form.
   a. $BaCl_2 + Na_2SO_4 \rightarrow BaSO_4 + 2NaCl$
      ionic equation: $Ba^{2+} + 2Cl^- + 2Na^+ + SO_4^{2-} \rightarrow BaSO_4 + 2Na^+ + 2Cl^-$
      net ionic equation: $Ba^{2+} + SO_4^{2-} \rightarrow BaSO_4$

   b. $Na_2CO_3 + 2HCl \rightarrow 2NaCl + CO_2 + H_2O$
      ionic equation: $2Na^+ + CO_3^{2-} + 2H^+ + 2Cl^- \rightarrow 2Na^+ + 2Cl^- + CO_2 + H_2O$
      net ionic equation: $CO_3^{2-} + 2H^+ \rightarrow CO_2 + H_2O$

   c. $NiS + 2HCl \rightarrow NiCl_2 + H_2S$
      ionic equation: $NiS + 2H^+ + 2Cl^- \rightarrow Ni^{2+} + 2Cl^- + H_2S$
      net ionic equation: $NiS + 2H^+ \rightarrow Ni^{2+} + H_2S$

5. Select the polyprotic acids from the following list.
   HCN, H_2CO_3, HCl, HF, H_2SO_3, H_3PO_4, H_2SeO_3, HBr
   H_2CO_3, H_2SO_3, H_3PO_4, H_2SeO_3

Write T for true or F for false. If a statement is false, replace the underlined word or phrase with one that will make the statement true, and write the correction on the blank provided.

F: neutralization 6. A salt is formed as a product in the <u>equilibrium</u> reaction of an acid and a base.

T 7. Salts may be formed from the reaction of acidic or basic anhydrides with a corresponding acid, base, or anhydride.

T 8. The percent ionization of a weak acid or base dissolved in water is a measure of the amount of <u>ionization</u> of the acid or base in solution.

Answer each question.

9. Write the equilibrium constant expression, K_{eq}, for the following acid ionization reaction.
   $HOBr + H_2O \rightleftharpoons H_3O^+ + OBr^-$

   $K_{eq} = \dfrac{[OBr^-][H_3O^+]}{[HOBr][H_2O]}$

10. Write the ionization constant expression, K_a, for the reaction in question 9.

    $K_a = \dfrac{[OBr^-][H_3O^+]}{[HOBr]}$

11. Calculate the hydronium ion concentration of a $0.25M$ solution of hypochlorous acid, HOCl, for which $K_a = 3.5 \times 10^{-8}$. The equation for ionization of this acid is $HOCl + H_2O \rightleftharpoons H_3O^+ + OCl^-$.

    $K_a = \dfrac{[OCl^-][H_3O^+]}{[HClO]} = 3.5 \times 10^{-8}$

    Let $[H_3O^+] = [OCl^-] = x$.
    $[HClO] = (0.25 - x)$

    $K_a = \dfrac{[OCl^-][H_3O^+]}{[HClO]} = \dfrac{(x)(x)}{(0.25 - x)} = 3.5 \times 10^{-8}$

    $K_a = \dfrac{x^2}{(0.25 - x)} = 3.5 \times 10^{-8}$

    Assuming x is $<< 0.25$, then $(0.25 - x) = 0.25$.

    $K_a = \dfrac{x^2}{0.25} = 3.5 \times 10^{-8}$

    $x^2 = (0.25)(3.5 \times 10^{-8}) = 8.8 \times 10^{-9}$
    $x = 9.4 \times 10^{-5} M$

12. Calculate the percent of ionization of $0.010M$ acetic acid solution, if the hydronium ion concentration is $4.2 \times 10^{-4} M$.

    percent of ionization $= \dfrac{4.2 \times 10^{-4} M}{0.010 M} \times 100\% = 4.2\%$

Chapter 24
STUDY GUIDE

24.1 WATER EQUILIBRIA

Complete the sentence or answer the question.

1. Define or explain the following. (Include formulation where applicable.)

   a. solubility product constant, K_{sp} of a saturated solution The solubility product constant, K_{sp}, of a saturated solution is derived from the equilibrium equation of the saturated solution. K_{sp} is a special case of equilibrium in which an ionic compound is in a solution so concentrated that the solid is in equilibrium with its ions. K_{sp} equals the product of the concentrations of the ions of the compound in the saturated solution.

   b. ion product constant of water, K_w The ion product constant of water, K_w, is derived from the equilibrium equation of the ionization of water and is constant for all dilute aqueous solutions at room temperature. K_w is found according to the following equation: $K_w = [H_3O^+][OH^-] = 1.00 \times 10^{-14}$

   c. pH pH is the measure of hydronium ion concentration, or acidity, of a solution.

   d. hydrolysis Hydrolysis is the reaction of a salt with water to form an acidic or basic solution.

   e. buffer solution A buffer solution can absorb moderate amounts of acids or bases without a significant change in the hydronium concentration, pH. A buffer provides ions that will react with and neutralize H_3O^+ or OH^- if they are introduced into the solution.

2. The addition of a common ion to a saturated solution decreases the solubility of a substance in solution.

3. Write the solubility product expression for each of the following.

   a. $BaSO_4$ $K_{sp} = [Ba^{2+}][SO_4^{2-}]$ d. $MgNH_4PO_4$ $K_{sp} = [Mg^{2+}][NH_4^+][PO_4^{3-}]$
   b. CaF_2 $K_{sp} = [Ca^{2+}][F^-]^2$ e. Ag_2CrO_4 $K_{sp} = [Ag^+]^2[CrO_4^{2-}]$
   c. Bi_2S_3 $K_{sp} = [Bi^{3+}]^2[S^{2-}]^3$

4. The concentration of water in pure water is 55.6 mol/dm^3.

5. Write the formula for calculating the pH from the concentration of H_3O^+.
   pH = $-\log[H_3O^+]$

6. Find the pH of solutions with the following H_3O^+ concentrations.

   a. $4.21 \times 10^{-4} M$
      pH = $-\log[4.21 \times 10^{-4}]$ = 3.38

   b. $5.00 \times 10^{-6} M$
      pH = $-\log[5.00 \times 10^{-6}]$ = 5.30

   c. $9.11 \times 10^{-8} M$
      pH = $-\log[9.11 \times 10^{-8}]$ = 7.04

   d. $2.14 \times 10^{-13} M$
      pH = $-\log[2.14 \times 10^{-13}]$ = 12.67

   e. $5.21 \times 10^{-10} M$
      pH = $-\log[5.21 \times 10^{-10}]$ = 9.28

   f. $1.45 \times 10^{-2} M$
      pH = $-\log[1.45 \times 10^{-2}]$ = 1.84

Match the acid-base reaction type with the examples given.

a. weak base + strong acid
b. strong acid + strong base
c. weak acid + strong base

b 7. $H^+ + OH^- \rightleftarrows H_2O$
c 8. $HC_2H_3O_2 + OH^- \rightleftarrows H_2O + C_2H_3O_2^-$
a 9. $NH_3 + H^+ \rightleftarrows NH_4^+$

10. Using the rules outlined in Chapter 24, classify each of the following solutions as acidic, basic, or neutral.

    a. $NaNO_3$ neutral   d. NH_4Br acidic
    b. KCN basic   e. $AlCl_3$ acidic
    c. Na_2CO_3 basic

Write T for true or F for false. If a statement is false, replace the underlined word or phrase with one that will make the statement true, and write your correction on the blank provided.

T _____ 11. Buffers are most efficient at neutralizing added acids or bases when the concentrations of HA and A⁻ (or MOH and M⁺) are the same.

F; hydronium _____ 12. The pH is the measure of oxygen in a solution.

F; poor _____ 13. Pure water is a good conductor of electricity.

T _____ 14. K_w is a constant for all dilute aqueous solutions at room temperature.

F; acidic _____ 15. As the hydronium concentration increases in a solution, the solution is made more basic.

Chapter 24
STUDY GUIDE
24.2 TITRATION

Complete the sentence or answer the question.

1. Define or explain the following.
   a. indicator a weak organic base or acid whose color differs from the color of its conjugate acid or base
   It is used to show the point of neutralization in a reaction.
   b. titration an analytical method in which a standard solution is used to determine the concentration of another solution
   c. standard solution a solution for which the concentration is known

2. Explain what a pH meter is and what it is used for.
   A pH meter is an electronic device that directly indicates the pH of a solution when its electrodes are immersed in the solution. It is used to determine the degree of acidity in a solution.

3. Explain how a titration is carried out.
   To carry out a titration, a buret is filled with a standard solution. A small amount of indicator is added to a measured amount of a solution of unknown concentration. The buret is opened and the standard solution is allowed to slowly flow into the solution to be titrated. When a color change occurs with a single drop, the end point, or neutralization, of the reaction is indicated.

4. A titration is performed in which 22.3 cm³ of 0.240M NaOH reacts with a 50.0-cm³ sample of $HC_2H_3O_2$. What is the concentration of the $HC_2H_3O_2$?
   The neutralization equation is $NaOH + HC_2H_3O_2 \rightarrow NaC_2H_3O_2 + H_2O$
   The concentration of NaOH that reacts is equal to the concentration of $HC_2H_3O_2$:

   $$22.3 \text{ cm}^3 \left| \frac{0.240 \text{ mol NaOH}}{1.00 \text{ dm}^3} \right| \frac{1 \text{ dm}^3}{1000 \text{ cm}^3} = 0.005\ 35 \text{ mol NaOH}$$

   0.005 35 moles NaOH = 0.005 35 moles $HC_2H_3O_2$

   $$\frac{0.005\ 35 \text{ moles } HC_2H_3O_2}{50.0 \text{ cm}^3} \left| \frac{1000 \text{ cm}^3}{1 \text{ dm}^3} \right. = 0.107M\ HC_2H_3O_2$$

Write T for true or F for false. If a statement is false, replace the underlined word or phrase with one that will make the statement true, and write your correction on the blank provided.

F; colorless ____ 5. Titration indicators are most useful when they are used on <u>colorful</u> solutions.

T ____ 6. If an acid is added to a base, a <u>neutralization reaction</u> occurs.

T ____ 7. Many indicators must be used in order to test for pH over a <u>wide</u> range of the pH scale.

Answer each question.

16. Label the parts of the pH scale in terms of acid, base, and neutral segments.

    pH = 0 ———— pH = 7 ————▶ pH = 14
    acid   neutral   base

17. What is the concentration of silver in a saturated solution of $AgC_2H_3O_2$? The K_{sp} of $AgC_2H_3O_2$ is 2.00×10^{-3}.
    The equilibrium equation for $AgC_2H_3O_2$ is $AgC_2H_3O_2 \rightleftharpoons Ag^+ + C_2H_3O_2^-$ and $K_{sp} = [Ag^+][C_2H_3O_2^-]$.
    Because the concentration of Ag^+ and $C_2H_3O_2^-$ are equal, let $[Ag^+] = x$, thus
    $K_{sp} = 2.00 \times 10^{-3} = x^2$
    $x = [Ag^+] = 0.045M$

18. Will a precipitate of $CaSO_4$ form in **hard water** if the Ca^{2+} concentration is 0.010M and the SO_4^{2-} concentration is 0.001 00M? Show your calculations. (K_{sp} of $CaSO_4 = 3.0 \times 10^{-5}$)
    If the actual concentration product of Ca^{2+} and SO_4^{2-} exceeds the solubility product constant for $CaSO_4$, then a precipitate will form.
    $[Ca^{2+}][SO_4^{2-}] = (0.010)(0.001\ 00) = 1.0 \times 10^{-5}$
    which is less than the solubility product constant of 3.0×10^{-5}; thus a precipitate will not form.

19. If 0.40 dm³ of 0.020M Na_2SO_4 is added to 0.80 dm³ of hard water, where $[Ca^{2+}] = 0.010M$ (as in question 18), will a precipitate form? Show your calculations.
    The total volume of solution is 0.40 dm³ + 0.80 dm³ = 1.2 dm³.

    $$[Ca^{2+}] = \frac{0.010M}{1} \left| \frac{0.80 \text{ dm}^3}{1.2 \text{ dm}^3} \right. = 0.0067M$$

    $$[SO_4^{2-}] = \frac{0.020M}{1} \left| \frac{0.40 \text{ dm}^3}{1.2 \text{ dm}^3} \right. = 0.067M$$

    $[Ca^{2+}][SO_4^{2-}] = (0.0067)(0.067) = 4.5 \times 10^{-5}$
    The concentration product is greater than the solubility product constant of 3.0×10^{-5}; thus a precipitate will form.

Chapter 25
STUDY GUIDE

25.1 OXIDATION AND REDUCTION PROCESSES

Complete each of the following.

1. A reaction in which ions or atoms undergo changes in electron structure is referred to as a(n) oxidation-reduction (redox) reaction.

2. The loss of electrons from an atom or ion is called oxidation.

3. The gain of electrons by an atom or ion is called reduction.

4. The substance in a redox reaction that undergoes oxidation is referred to as a(n) reducing agent.

5. The substance in a redox reaction that undergoes reduction is referred to as a(n) oxidizing agent.

6. In the reaction that follows, H_2 is oxidized and Cl_2 is reduced. $H_2 + Cl_2 \rightarrow 2HCl$

7. The oxidation number of a free element is zero.

8. In most reactions, the oxidation number of the calcium ion is 2+.

9. The oxidation number of the oxide ion is generally 2−.

10. The sum of the oxidation numbers of all of the atoms in HNO_3 is zero.

11. The oxidation number of N in $Mg(NO_3)_2$ is 5+.

12. The oxidation number of S in SCl_2 is 2+.

13. In the reaction that follows, the oxidizing agent is H_2O. $C + H_2O \rightarrow CO + H_2$

14. In the reaction given in question 12, the element being oxidized is C.

15. In the reaction given in question 12, the total number of electrons transferred (lost or gained) is 2.

16. The oxidation number of P in PO_4^{3-} is 5+.

Write true (T) or false (F) for each of the following.

F a. The following is an example of a redox reaction: $HCl + NaOH \rightarrow NaCl + H_2O$

T b. The term *oxidation* was first applied to the combining of oxygen with other elements.

T c. In an oxidation-reduction reaction, the number of electrons lost must equal the number gained.

F d. If a substance gains electrons readily, it is said to be a strong reducing agent.

T e. The oxidation number of a monatomic ion is equal to the charge on the ion.

T f. In general, ions of the Group 17 elements have oxidation numbers of 1−.

F g. The oxidizing agent in the following reaction is Na. $2Na + S \rightarrow Na_2S$

T h. A total of 2 electrons are transferred in the reaction given in question 16g.

T i. The element that undergoes reduction in the reaction given in question 16g is S.

17. Complete the table below for each reaction specified.

Reaction	Element oxidized	Element reduced	Oxidizing agent	Reducing agent	Total e⁻ transferred
$2HBr + Cl_2 \rightarrow 2HCl + Br_2$	Br	Cl	Cl_2	HBr	2
$Zn + I_2 \rightarrow ZnI_2$	Zn	I	I_2	Zn	2
$Fe_2O_3 + 3CO \rightarrow 2Fe + 3CO_2$	C	Fe	Fe_2O_3	CO	6
$16H^+ + 2MnO_4^- + 5C_2O_4^{2-} \rightarrow 2Mn^{2+} + 8H_2O + 10CO_2$	C	Mn	MnO_4^-	$C_2O_4^{2-}$	10

Chapter 25
STUDY GUIDE

25.2 BALANCING REDOX EQUATIONS

1. Complete the steps outlined below in order to balance the following oxidation-reduction reaction:
$HNO_3 + S \rightarrow NO_2 + H_2SO_4 + H_2O$

   a. Write the skeleton equation for the reaction.
   $NO_3^- + S \rightarrow NO_2 + HSO_4^-$

   b. Assign oxidation numbers to all atoms involved in the reaction.
   $\ 5+\ 2-\ 0 \ 4+\ 2-\ 1+\ 6+\ 2-$
   $NO_3^- + S \rightarrow NO_2 + HSO_4^-$

   c. Identify the substance being oxidized, and then write and balance the oxidation reaction in terms of both atoms and charge. In acidic solutions, it may be necessary to add H^+ ions and H_2O molecules to balance the reaction. In basic solutions, it may be necessary to add a sufficient number of OH^- ions to each side of the equation to combine with any excess H^+ ions and form H_2O molecules.

   Oxidation: $\qquad S + 4H_2O \rightarrow HSO_4^- + 7H^+ + 6e^-$

   d. Identify the substance being reduced, and then write and balance the reduction reaction in terms of both atoms and charge.
   Reduction: $NO_3^- + e^- + 2H^+ \rightarrow NO_2 + H_2O$

   e. Combine the two half-reactions, first multiplying one or both by the factor(s) needed to balance the electron transfer. Cancel the electrons, as well as the excess number of any species that appear on both sides of the equation, from the final reaction.

   $S + 4H_2O \rightarrow HSO_4^- + 6e^- + 7H^+$
   $6[NO_3^- + e^- + 2H^+ \rightarrow NO_2 + H_2O]$
   combined: $S + 4H_2O + 6NO_3^- + \cancel{6e^-} + \cancel{12H^+} \rightarrow HSO_4^- + \cancel{6e^-} + \cancel{7H^+}^{\ 5} + 6NO_2 + 6H_2O^{\ 2}$
   Net: $S + 6NO_3^- + 5H^+ \rightarrow HSO_4^- + 6NO_2 + 2H_2O$

   f. Perform a final check on the balanced equation to ensure that both atoms and charge are balanced.
   Atoms and charge are balanced.

Apply the steps given above to balance each of the following oxidation-reduction reactions.

2. $Br_2 + SO_2 + H_2O \rightarrow H_2SO_4 + HBr$
   $Br_2 + SO_2 + 2H_2O \rightarrow 2Br^- + HSO_4^- + 3H^+$

3. $PbS + H_2O_2 \rightarrow PbSO_4 + H_2O$
   $PbS + 4H_2O_2 \rightarrow PbSO_4 + 4H_2O$

4. $H_3AsO_4 + Zn \rightarrow AsH_3 + Zn^{2+}$
   $Zn + H_3AsO_4 + 5H^+ \rightarrow As^{3+} + 4H_2O + Zn^{2+}$

5. $PH_3 + O_2 \rightarrow P_4O_{10} + H_2O$
   $4P^{3+} + 6H_2O + 2O_2 \rightarrow P_4O_{10} + 12H^+$

6. $NO_2 + OH^- \rightarrow NO_2^- + NO_3^-$ (in basic solution)
   $2NO_2 + 2OH^- \rightarrow NO_2^- + NO_3^- + H_2O$

7. $K_2Cr_2O_7 + Na_2SO_3 + HCl \rightarrow Cr_2(SO_4)_3 + KCl + NaCl + H_2O$
   $Cr_2O_7^{2-} + 8H^+ + 3SO_3^{2-} \rightarrow 2Cr^{3+} + 4H_2O + 3SO_4^{2-}$

8. $Cu(OH)_2 + HPO_3^{2-} \rightarrow Cu_2O + PO_4^{3-}$
   $2Cu(OH)_2 + HPO_3^{2-} \rightarrow Cu_2O + PO_4^{3-} + 2H_2O + H^+$

9. $TeO_2 + BrO_3^- \rightarrow H_6TeO_6 + Br_2$
   $5TeO_2 + 2BrO_3^- + 14H_2O + 2H^+ \rightarrow 5H_6TeO_6 + Br_2$

10. $Bi_2S_3 + NO_3^- \rightarrow Bi^{3+} + NO + S$
    $Bi_2S_3 + 2NO_3^- + 8H^+ \rightarrow 2Bi^{3+} + 2NO + 3S + 4H_2O$

11. $NO + H_5IO_6 \rightarrow NO_3^- + IO_3^-$
    $2NO + 3H_5IO_6 \rightarrow 2NO_3^- + 3IO_3^- + 5H_2O + 5H^+$

Chapter 26
STUDY GUIDE

26.1 CELLS

1. Indicate if the bulb shown in Figure 1 will light or not by placing a plus(+) or minus(−) sign, respectively, in the blank to the left of each of the following situations.

   − a. The electrodes are connected with a piece of glass.
   − b. The beaker is filled with air.
   − c. The beaker is filled with an electrolyte.
   + d. The electrodes are connected by a wire.
   − e. The beaker is filled with alcohol.

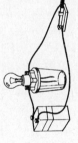

Figure 1

2. In situation c of question 1, write the type of conduction, if any, of the materials.

   a. wires metallic conduction   c. bulb filament metallic conduction
   b. electrolyte electrolytic conduction   d. electrodes metallic conduction

Write T for true or F for false. If a statement is false, replace the underlined word or phrase with one that will make the statement true, and write your correction on the blank provided.

T   3. A galvanometer is an instrument that detects electric current.
F: volts   4. Potential difference is measured in units of amperes.
F: a   5. Current flows through wires in which there is no potential difference between the ends.
T   6. Metallic conduction takes place because the electrons of metals are free to move when a small potential difference is applied.
F: ions   7. Electrolytic conduction takes place because electrons move freely in aqueous solutions.

Complete the sentence.

8. An electrolyte is any substance that produces ions in solution.
9. The process by which an electric current produces a chemical change is called electrolysis .
10. In a cell, oxidation and reduction occur as separate half-reactions at the same time .
11. A device that converts chemical potential energy into electric energy is called a voltaic cell .
12. The zinc-copper cell reaction, in which Zn is oxidized to Zn^{2+} and Cu^{2+} is reduced to Cu, is represented as $Zn|Zn^{2+}||Cu^{2+}|Cu$.

Using the letters of the terms below, label each of the diagrams in Figure 2.

a. anode (Pb)   g. cathode (steel)
b. anode (Zn can)   h. dry cell
c. anode (Zn shell)   i. electrolyte (H_2SO_4)
d. automobile battery   j. electrolyte (KOH and paste of $Zn(OH)_2$ and HgO)
e. cathode (graphite)   k. electrolyte (paste of MnO_2, NH_4Cl, and powdered graphite)
f. cathode (PbO_2)   l. mercury battery

Figure 2

13. e
14. k
15. c
16. h
17. f
18. a
19. i
20. d
21. g
22. b
23. j
24. l

Chapter 26
STUDY GUIDE

26.2 QUANTITATIVE ELECTROCHEMISTRY

Write T for true or F for false. If a statement is false, replace the underlined word or phrase with one that will make the statement true, and write your correction on the blank provided.

F; electrons 1. The difference between two half-cells is a measure of the relative tendency of the two substances to take on <u>protons</u>.

F; reductions 2. The half-reactions listed in Table 26.1 of the text are written as <u>oxidations</u>.

T 3. The symbol for standard reduction potential (25°C, 101.325 kPa, 1M) is <u>E^0</u>.

F; hydrogen 4. The half-reaction of <u>carbon</u> is assigned a reduction potential of 0.000 0 V.

T 5. The standard reduction potential of a substance is an <u>intensive</u> property.

Use Table 26.1 in the text to answer questions 6-11.

6. Rank the following substances in order of decreasing reduction strength.
Ag^+ Au^+ Cs^+ K^+ Li^+ Na^+

   Li^+ K^+ Na^+ Cs^+ Ag^+ Au^+

7. Explain why the reaction, $Cu(cr) + 2Ag^+(aq) \rightarrow 2Ag(cr) + Cu^{2+}(aq)$ will occur spontaneously.

   Because the E^0 of silver ions is more positive than that of copper ions, the silver ions have a greater attraction for electrons than copper. Therefore the oxidation of copper and the reduction of Ag^+ will take place spontaneously.

8. Predict if each of the following reactions will occur spontaneously as written.

   a. $Cu(cr) + Pd^{2+}(aq) \rightarrow Cu^{2+}(aq) + Pd(cr)$

   $Cu \rightarrow Cu^{2+} + 2e^-$ $E^0 = -0.340$ V

   $Pd^{2+} + 2e^- \rightarrow Pd$ $E^0 = +0.915$ V

   Adding E^0 gives a positive potential. Therefore a reaction occurs.

   b. $2Ag(cr) + Co^{2+}(aq) \rightarrow Co(cr) + 2Ag^+(aq)$

   $Ag \rightarrow Ag^+ + e^-$ $E^0 = -0.7991$ V

   $Co^{2+} + 2e^- \rightarrow Co$ $E^0 = -0.277$ V

   The sum is negative. Therefore no reaction occurs.

9. Consider the following cell (standard conditions).
   $Zn\,|\,Zn^{2+}\,||\,Ni^{2+}\,|\,Ni$

   a. What substance is oxidized in the cell?

   Zn

   b. Write the oxidation half-cell reaction.

   $Zn \rightarrow Zn^{2+} + 2e^-$

   c. What is the potential for the oxidation half-cell reaction?

   +0.7626 V

   d. What substance is reduced in the cell?

   Ni^{2+}

   e. What is the reduction half-cell reaction?

   $Ni^{2+} + 2e^- \rightarrow Ni$

   f. What is the potential for the reduction half-cell reaction?

   −0.257 V

   g. What is the potential produced by the cell?

   +0.506 V

10. Predict the potential for each of the following cells.

    a. $Zn\,|\,Zn^{2+}\,||\,Cu^{2+}\,|\,Cu$

    $Zn \rightarrow Zn^{2+} + 2e^- + 0.7626$ V

    $Cu^{2+} + 2e^- \rightarrow Cu + 0.340$ V

    0.7626 V + 0.340 V = 1.103 V

    b. $Fe\,|\,Fe^{2+}\,||\,Ag^+\,|\,Ag$

    $Fe \rightarrow Fe^{2+} + 2e^- + 0.44$ V

    $Ag^+ + e^- \rightarrow Ag + 0.7991$ V

    0.44 V + 0.7991 V = 1.24 V

11. The relationship $E = -0.05916$ pH indicates that E is a __linear__ function of pH.

12. The following cell has a standard potential of 0.459V.
    $Cu\,|\,Cu^{2+}\,||\,Ag^+\,|\,Ag$
    Write the chemical reaction for the cell.

    $Cu(cr) + 2Ag^+(aq) \rightarrow Cu^{2+}(aq) + 2Ag(cr)$

Chapter 27 STUDY GUIDE

27.1 INTRODUCTORY THERMODYNAMICS

1. For each of the following processes, determine whether the indicated quantities of the system are greater than zero, equal to zero, or less than zero. Write your answer in the blank following each quantity. If not enough information is given to determine a value, place an X in the blank.

   a. A system loses heat.
      q <0   w   X ΔU X

   b. A sample of a gas is compressed without heat transfer.
      q =0   w   >0 ΔU >0

   c. A system does work without a decrease in its internal energy.
      q >0   w   <0 ΔU =0

   d. A sample of a gas is heated. As the gas expands, it does work pushing a piston.
      q >0   w   <0 ΔU X

   e. The internal energy of a confined gas increases without work being done.
      q >0   w   0 ΔU >0

Figure 1 is a pressure-temperature graph of a system showing two paths (arrows) that represent independent ways of changing the system. The internal energy of the system is given by $\Delta U = q + w$. Use Figure 1 to answer questions 2–9.

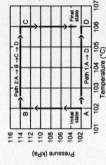

Figure 1

2. Underline each of the following variables that is a state function.
   P   T   q   w   <u>ΔU</u>

3. Underline each of the following paths that represents an isobaric process.
   A→B   <u>B→C</u>   C→D   <u>A→D</u>

4. Underline each of the following paths that represents an isothermal process.
   <u>A→B</u>   B→C   <u>C→D</u>   A→D

5. What is the final temperature of the system?
   106°C

Chapter 27 STUDY GUIDE

27.2 DRIVING CHEMICAL REACTIONS

For each of the following situations determine if the entropy change will be positive (+) or negative (−). Write a plus or minus sign in the blank provided.

+ 1. Calcium chloride crystals are pulverized in a mortar and pestle.
+ 2. Turpentine evaporates.
− 3. A mixture of iron and sulfur is separated using a magnet.
+ 4. A spark plug ignites a fuel-air mixture in an engine.
+ 5. Turpentine dissolves a paint stain.
− 6. Melted solder solidifies.
+ 7. Air is heated.
− 8. Graphite is recrystallized as diamond in a press.

9. Complete the table by naming each state function and giving the appropriate SI unit for a molar quantity.

Symbol	State function	SI unit
ΔH_f°	enthalpy of formation	J/mol
ΔG_f°	Gibbs free energy of formation	J/mol
ΔS°	entropy	J/mol·K
E°	standard cell potential	V

Answer the following questions.

10. What are standard states of temperature and pressure for measuring thermodynamics quantities?
    $T = 298.15$ K, $P = 100.000$ kPa

11. What symbol is used to indicate that a quantity has been measured under standard states?
    The quantity is followed by a superscript "°".

12. What is characteristic about the Gibbs free energies of reactions that are spontaneous?
    The values of the Gibbs free energies of the reactions are less than zero.

13. What is characteristic about the Gibbs free energies of reactions that are at equilibrium?
    The values of the Gibbs free energies of the reactions are equal to zero.

14. What is characteristic about the Gibbs free energies of reactions that are not spontaneous?
    The values of the Gibbs free energies of the reactions are greater than zero.

---

6. What is the final pressure of the system?
   102 kPa

7. What is ΔT of the system?
   4°C

8. What is ΔP of the system?
   0 kPa

9. If ΔH of the system is 20.0 J, determine q_p.
   $q_p = \Delta H = +20.0$ J

Figure 2 represents the changes of enthalpy of the reactions
$H_2(g) + \frac{1}{2}O_2(g) \rightarrow H_2O(l)$
$H_2O(l) \rightarrow H_2O(g)$

Use Figure 2 to answer questions 10 - 14.

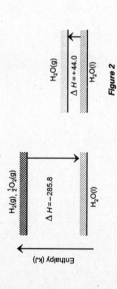

Figure 2

10. Which of the two reactions represents only a physical change?
    $H_2O(l) \rightarrow H_2O(g)$

11. Which reactants have enthalpies of formation of 0 kJ/mol?
    $H_2(g)$, $O_2(g)$

12. Which reactions are exothermic?
    $H_2(g) + \frac{1}{2}O_2(g) \rightarrow H_2O(l)$ and $H_2(g) + \frac{1}{2}O_2(g) \rightarrow H_2O(g)$

13. Write the chemical equation for the formation of $H_2O(g)$ from its elements.
    $H_2(g) + \frac{1}{2}O_2(g) \rightarrow H_2O(g)$

14. Determine the enthalpy of formation of $H_2O(g)$ from its elements.
    $\Delta H = \Delta H_1 + \Delta H_2 = (-285.8$ kJ$) + (+44.0$ kJ$) = -241.8$ kJ

Chapter 28
STUDY GUIDE

28.1 NUCLEAR STRUCTURE AND STABILITY

Answer each question.

1. Name two devices that produce high-energy particles used to bombard nuclei in the investigation of nuclear structure, and compare how the devices are used.
   An accelerator is used to increase velocity of specific charged particles. A reactor is used to expose nuclei to random bombardment by neutrons.

2. List three things that may happen when a nucleus is bombarded by a high-energy particle, and describe the possible effect of each on the bombarded nucleus.
   A particle may bounce off the nucleus; the nucleus may become metastable and give off energy. A particle may shatter the nucleus; nucleons are given off. A particle may merge with the nucleus; a new nucleus is formed.

3. On what three things does the type of change of a bombarded nucleus depend?
   the relative stability of the target nucleus; the kind of bombarding particle used; the energy of the bombarding particles

4. State two problems associated with the use of nuclear reactors to produce energy for practical use.
   Answers may include: disposal of radioactive waste materials and providing of adequate safety at reactor sites.

5. Radioactive element X decays to element Y. A sample initially contains 72 g of element X. At the end of three half-lives, how many grams of element X does the sample contain?
   9 g

Write T for true or F for false. If a statement is false, replace the underlined word or phrase with one that will make the statement true, and write your correction on the blank provided.

F; magnets   6. In a synchrotron, <u>drift tubes</u> are used to confine fast moving particles to the ring.

F; linear   7. In a <u>circular</u> accelerator, electromagnetic waves are used to accelerate particles to optimum speed.

F; no charge 8. Neutrons easily penetrate and can be absorbed by nuclei because the neutrons have <u>a negative charge</u>.

T   9. <u>Fission</u> involves the breaking apart of a heavy nucleus into two parts of about the same mass.

F; neutrons 10. The emission and absorption of <u>alpha particles</u> are necessary in sustaining a chain reaction.

T   11. In nuclear reactions, the law of conservation of <u>mass-energy</u> is upheld.

F; control rod 12. In a nuclear reactor, a <u>moderator</u> is a device that regulates the rate at which the chain reaction takes place.

F; decay   13. The half-life of a radioactive element is the amount of time it takes for half the atoms of a sample to <u>react with other atoms</u>.

Place a check mark (✓) in the blank next to each set of conditions that tends to favor spontaneous reactions. Place an X in the blank if the conditions do not favor spontaneous reactions.

	ΔH	T	ΔS			ΔH	T	ΔS
✓ 15.	<0	low	<0	X	19.	>0	low	<0
✓ 16.	<0	low	>0	X	20.	>0	low	>0
X 17.	<0	high	<0	X	21.	>0	high	<0
✓ 18.	<0	high	>0	✓	22.	>0	high	>0

23. Use the information in Table 1 to complete Table 2 for the following reaction that takes place under standard measurement conditions.

$$CaCO_3(cr) \rightarrow CaO(cr) + CO_2(g)$$

Table 1

State function	Compound		
	$CaCO_3(cr)$	$CaO(cr)$	$CO_2(g)$
ΔG_f° (kJ/mol)	−1.129	−0.641	−1.650
S° (J/mol·K)	92.8	39.7	885.6

Table 2

Quantity	Value
$\Sigma \Delta G_f^\circ$ (products)	−2.291 kJ
$\Sigma \Delta G_f^\circ$ (reactants)	−1.129 kJ
ΔG_r°	−1.162 kJ

Quantity	Value
ΣS_f° (products)	925.3 J/K
ΣS_f° (reactants)	92.8 J/K
S_r°	832.5 J/K

24. Is the reaction in question 23 spontaneous? Explain your reasoning.
    yes; ΔG_r is negative.

Name _____ Date _____ Class _____

F: 2.5 g   14. A sample of a radioactive substance has a mass of 10 g. At the end of two half-lives, 2.5 g of the original sample remains.

F: moderator   15. Because it does not absorb neutrons, water is often used as a coolant in nuclear reactors.

F: nuclear reactor   16. A synchrotron is a device for controlling fission.

T   17. The containment vessel of a nuclear reactor protects people from the heat and radiation of the reactor core.

Match each substance with the part of a nuclear reactor in which the substance is used. Some substances may be used in more than one part of the reactor.

a. moderator
b. control rods
c. coolant
d. fuel
e. containment vessel

a, c   18. water
d   19. uranium-235
b   20. cadmium
d   21. plutonium-239
c   22. molten sodium
c   23. helium
a   24. graphite
b   25. boron

26. Complete the table below to show how atomic number and mass number are affected in various transmutations.

Decay mode	Effect on atomic number	Effect on mass number
α-particle	decreases by 2	decreases by 4
β-particle	increases by 1	remains the same
K-capture	decreases by 1	remains the same

27. Label the parts of the nuclear reactor shown in the figure and describe the function of each part.

a. control rods: regulate fission rate
b. containment vessel: shields reactor walls and protects people from heat and radiation
c. moderator: slows down fission neutrons
d. coolant: absorbs heat produced by fission
e. fuel: fissionable material that supplies neutrons

Use Figure 28.6 in your textbook to answer the following questions about the nuclear decay of uranium-238 to lead-206.

28. What is the total number of alpha particles emitted during the process?
8

29. What is the total number of beta particles emitted?
6

30. What is the atomic number of the nuclide formed at the end of step 5?
88

31. What is the mass number of the nuclide formed at the end of step 8?
214

32. What is the name of the nuclide formed at the end of step 9?
bismuth-214

33. What is the name of the nuclide with the shortest half-life?
polonium-214

34. How is the nuclide produced in step 11 different from the nuclide produced at the end of the decay series?
The isotope of lead at step 11 (Pb-210) has 128 neutrons and is unstable; Pb-206, formed at the end, has 124 neutrons and is stable.

Chapter 28
STUDY GUIDE

28.2 NUCLEAR APPLICATIONS

Write T for true or F for false. If a statement is false, replace the underlined word or phrase with one that will make the statement true, and write your correction on the blank provided.

F: transuranium   1. Elements that have atomic numbers greater than 92 are known as the <u>transmutation</u> elements.

F: alpha particles   2. Because of their relatively large mass and charge, the particles that, once inside the body, are most damaging to cells are <u>beta particles</u>.

T   3. Alpha particles are the <u>least</u> penetrating type of radiation.

F: detected   4. Radioactive elements are useful as tracers because they can be easily <u>disintegrated</u>.

F: fusion   5. In <u>fission</u>, the nuclei of small atoms unite to form a larger nucleus.

T   6. Fusion generally produces a <u>greater</u> amount of energy per particle than does fission.

F: cloth or wood   7. Carbon-14 dating is most useful for determining the age of <u>rocks</u>.

F: absorbed   8. Radiation damage to living cells is indicated by the amount of <u>emitted</u> radiation.

F: higher   9. Transmutations resulting from "packing" neutrons into the nuclei of certain elements produce elements with atomic numbers <u>lower</u> than those of the original elements.

Answer each question.

10. List three ways in which transuranium elements may be produced.

    by packing a nucleus with neutrons, by capture of neutrons produced by a nuclear explosion, and by bombarding a target nucleus with other elements

11. Differentiate between a gray and a sievert.

    A gray measures the transfer of radiation energy to living tissue; a sievert measures the dose of radiation absorbed by living tissue.

12. List four sources of radiation to which humans are generally exposed.

    Answers may include: naturally occurring radioactive elements found in rocks, cosmic radiation, radioactive elements occurring naturally in food and water, and radiation from testing of nuclear weapons and from nuclear-power-plant operation.

13. List four ways in which radioactive nuclides are put to practical use.

    Answers may include: to date rocks, fossils, and artifacts; to study reaction mechanisms; in quantitative analysis; and as biological tracers.

14. State three problems associated with the development of plasma fusion reactors.

    Answers may include: producing the high-energy input required to achieve necessary high temperatures, finding a suitable material or way to contain high-temperature plasma, and counteracting the electric field generated by high-velocity plasma.

Chapter 29
STUDY GUIDE

29.1 HYDROCARBONS

Complete each sentence.

1. A chain compound in which all carbon-carbon bonds are single is called an   alkane   or a   saturated hydrocarbon .

2. Each alkane differs from the next by a(n) CH_2 group.

3. With increasing molecular mass of compounds within a homologous series, the boiling point   increases .

Write the name that corresponds with the following chemical formulas of compounds or radicals.

4. C_5H_{11} pentyl
5. C_8H_{18} octane
6. C_3H_8 propane
7. C_4H_9 butyl

Write T for true or F for false. If a statement is false, replace the underlined word or phrase with one that will make the statement true, and write your correction on the blank provided.

F: an unbranched   8. In <u>a branched</u> chain molecule, the numbering of carbon atoms can begin at either end of the chain.

F: alphabetical   9. When a compound has more than one branch, radicals appear in <u>numerical order</u> in the name of the compound.

T   10. Prefixes are used in the naming of compounds in which two or more substituent groups are alike.

F: three one-carbon branches   11. The term *trimethyl* means <u>one three-carbon branch</u>.

T   12. Isomers <u>are not</u> named according to the total number of carbon atoms in a molecule.

Match the general formulas with the corresponding class of organic compounds.

a. ester
b. ether
c. ketone
d. acid
e. alcohol

d   13. $R-\underset{\underset{O}{\|}}{C}-O-H$

a   14. $R-\underset{\underset{O}{\|}}{C}-O-R$

e   15. $R-O-H$

c   16. $R-\underset{\underset{O}{\|}}{C}-R$

b   17. $R-O-R$

Name _____ Date _____ Class _____

Define the following terms.

18. cycloalkanes cyclic forms of saturated hydrocarbons _____

19. alkenes unsaturated hydrocarbons with a double bond between carbon atoms _____

20. unsaturated hydrocarbons hydrocarbons that contain multiple bonds _____

21. alkynes molecules that contain a triple bond between carbon atoms _____

22. aromatic hydrocarbons compounds derived from benzene, often having distinctive odors __

23. How does a benzene ring differ from cyclohexane?
Cyclohexane is an alkane having the formula C_6H_{12} composed of six single-bonded carbon atoms, whereas benzene is an unsaturated hydrocarbon having the formula C_6H_6 in which carbon atoms are bonded by a system of delocalized electrons equivalent to alternating double bonds.

Draw the structural formulas for the following compounds in the spaces provided.

24. benzene

25. cyclohexane

26. phenyl radical

27. methylcyclohexane

28. 1,3-dimethylcyclopentane

Name _____ Date _____ Class _____

29. toluene

30. 2,2,4-trimethylpentane

$$CH_3-CH-CH_2-\underset{\underset{CH_3}{|}}{\overset{\overset{CH_3}{|}}{C}}-CH_3$$
$$\underset{CH_3}{|}$$

31. 1-methyl-3-ethylcyclohexane

Answer each of the following.

32. What do benzene, toluene, and xylene have in common?
They are all aromatic, and all are synthesized from petroleum.

33. What materials is styrene used to make?
synthetic rubber, plastics, and paints

Chapter 29
STUDY GUIDE

29.2 OTHER ORGANIC COMPOUNDS

Match the corresponding terms by choosing from the list below.

a. chloromethane
b. methanol
c. 1,2-dichloroethane
d. chloroethene
e. trichloromethane
f. 1,2-ethanediol

- e 1. is also known as chloroform
- a 2. has the formula CH_3Cl
- d 3. is also known as vinyl chloride
- c 4. is also known as ethylene dichloride
- b 5. has the formula CH_3OH
- f 6. is also known as ethylene glycol

Write T for true or F for false. If a statement is false, replace the underlined word or phrase with one that will make the statement true, and write your correction on the blank provided.

- F; phenols 7. Because aromatic hydroxyl compounds have properties that differ somewhat from those of most alcohols, they are classified as <u>bases</u>.
- F; an anesthetic 8. Ethoxyethane was used for many years as <u>an antiseptic</u>.
- T 9. Alcohols are <u>neither acidic or basic</u>.
- F; are not soluble 10. Alcohols with four or more carbon atoms <u>are soluble</u> in water.

Fill in the missing term(s).

11. The most important ketone in industry is __acetone (or propanone)__.
12. Methanal is more commonly known as __formaldehyde__.
13. Both aldehydes and ketones are characterized by the presence of a(n) __carbonyl__ group.
14. The effect of one functional group on another is called a(n) __inductive effect__.
15. Because they do not ionize greatly in water, most organic acids are __weak__ acids.
16. The __carboxylic__ acid group characterizes most organic acids.
17. Acetic anhydride is made from __acetic acid__ by the removal of a molecule of water.
18. In the reaction between ethanol and acetic acid, __ethyl acetate__ and __water__ are produced.

Define the following terms.

19. amine   <u>organic compound in which a nitrogen atom is bonded to alkyl groups and hydrogen atoms</u>

20. primary amine   <u>amine that has only one hydrogen atom replaced by an alkyl group</u>

21. amide   <u>organic compound characterized by a carbonyl group and an amine group</u>

22. nitrile   <u>organic compound characterized by a carbon-nitrogen triple bond</u>

Write the general formula for each of the following compounds.

23. nitro compounds

    $R-NO_2$

24. amines

    $R-NH_2$

25. nitriles

    $R-C\equiv N$

Chapter 30 STUDY GUIDE

30.1 ORGANIC REACTIONS

Match the reactants in questions 1-6 with the following products. Then identify the type of organic reaction illustrated in each question.

a. H-C(H)(H)-C₆H₅-H + HCl

b. H-C(H)(H)-C(H)(H)-OH

c. H-C(OH)(H)-C(H)(H)-C(H)(H)-C(H)(H)-H + H₂O

d. H-C(H)(H)-C(H)(H)-C(H)(H)-H

e. H-C(H)(H)-C=C(H)(H) + H₂O

f. C₆H₅-H + CH₃Cl

g. (chain of —C(H)(Cl)—C(H)(H)— repeating units)

__d; addition__ 1. H-C(H)=C(H)-H + HI →

__a; substitution__ 2. C₆H₆ + H-C(H)(H)-Cl →(AlCl₃)

__e; elimination__ 3. H-C(H)(H)-C(H)(OH)-C(H)(H)-H →(H₂SO₄)

__b; addition__ 4. H-C(H)=C(H)-H + H₂O →(dilute H₂SO₄)

__c; esterification__ 5. H-C(=O)-OH + H-C(H)(H)-OH →

__g; polymerization__ 6. H-C(H)=C(H)-Cl + H-C(H)=C(H)-Cl + ... →

Write T for true or F for false. If a statement is false, replace the underlined word or phrase with one that will make the statement true, and write your correction on the blank provided.

__F; substitution__ 7. In a condensation polymerization reaction, one functional group may be replaced by another.

__F; broken__ 8. In an addition reaction, a double bond between carbons is created.

__F; elimination__ 9. A reaction in which water is removed from an alcohol is an esterification reaction.

__F; saponification__ 10. Soap is made by the polymerization of fatty acids.

__T__ 11. Like molecules called monomers chain together to form polymers.

__F; elastomer__ 12. When an ester is deformed by an outside force, it returns to its original shape when the force is removed.

__T__ 13. Rayon is made from reconstituted cellulose.

__F; substitution__ 14. Many compounds containing benzene readily take part in addition reactions.

__T__ 15. In the fractional distillation of petroleum, the boiling point of the gasoline fractions must be lower than the boiling point of the fuel oil fractions.

__F; gasoline__ 16. Cracking is a process used to increase the yield of natural rubber.

__F; thermosetting__ 17. A thermoplastic material hardens upon heating.

__T__ 18. The higher the octane rating, the greater the percent of the fuel that burns evenly.

Chapter 30
STUDY GUIDE

30.2 BIOCHEMISTRY

Write T for true or F for false. If a statement is false, replace the underlined word or phrase with one that will make the statement true, and write your correction on the blank provided.

F; protein _____ 1. Enzymes are <u>lipid</u>-based biological catalysts.

F; amino _____ 2. Proteins are composed of <u>nucleic acids</u> linked by peptide bonds.

F; lipids _____ 3. Steroids, some vitamins, and fats are classified as <u>carbohydrates</u>.

T _____ 4. <u>Carbohydrates</u> are more soluble in water than in nonpolar solvents.

F; nucleic acids _____ 5. Protein synthesis in the human body is controlled by <u>amino acids</u>.

F; condensation _____ 6. Polysaccharides are formed by <u>addition</u> polymerization.

F; collagen _____ 7. Cartilage and tendons are composed mainly of <u>cellulose</u>.

Answer the following questions.

8. State three ways in which proteins may differ from one another.
   sequence of amino acids; way in which chain is coiled, folded, or twisted; type of bonding that shapes the
   polymer

9. List three glucose-based polysaccharides found in living things.
   glycogen, cellulose, starch

10. State three problems associated with the use of biomaterials.
    The body tends to reject foreign substances; biomaterials tend to cause blood clots; biomaterials tend to
    form fibrous tissue that may harden.

11. List four uses of biomaterials.
    in artificial skin, artificial blood vessels, artificial heart valves, and dialysis and plasmapheresis membranes

12. Compare the chemical composition of RNA with that of DNA.
    Both contain adenine, cytosine, and guanine as nitrogen bases; RNA contains the nitrogen base uracil; DNA
    contains the nitrogen base thymine; RNA is composed of ribose sugar; DNA is composed of deoxyribose
    sugar.

Refer to Table 30.1 in your textbook to answer the following questions.

13. Draw a structural formula for the amino acid serine.

$$NH_2-\overset{\overset{H}{|}}{\underset{\underset{CH_2}{|}}{C}}-\overset{O}{\overset{||}{C}}-OH$$
$$OH$$

14. Draw a structural formula for the amino acid glycine.

$$NH_2-\overset{\overset{H}{|}}{\underset{\underset{H}{|}}{C}}-\overset{O}{\overset{||}{C}}-OH$$

15. Use structural formulas to show how serine and glycine may combine to form two different dipeptides.

$$NH_2-\overset{H}{\underset{CH_2OH}{C}}-\overset{O}{\overset{||}{C}}-OH + NH_2-\overset{H}{\underset{H}{C}}-\overset{O}{\overset{||}{C}}-OH \rightarrow NH_2-\overset{H}{\underset{CH_2OH}{C}}-\overset{O}{\overset{||}{C}}-NH-\overset{H}{\underset{H}{C}}-\overset{O}{\overset{||}{C}}-OH + H_2O$$

$$NH_2-\overset{H}{\underset{H}{C}}-\overset{O}{\overset{||}{C}}-OH + NH_2-\overset{H}{\underset{CH_2OH}{C}}-\overset{O}{\overset{||}{C}}-OH \rightarrow NH_2-\overset{H}{\underset{H}{C}}-\overset{O}{\overset{||}{C}}-HN-\overset{H}{\underset{CH_2OH}{C}}-\overset{O}{\overset{||}{C}}-OH + H_2O$$